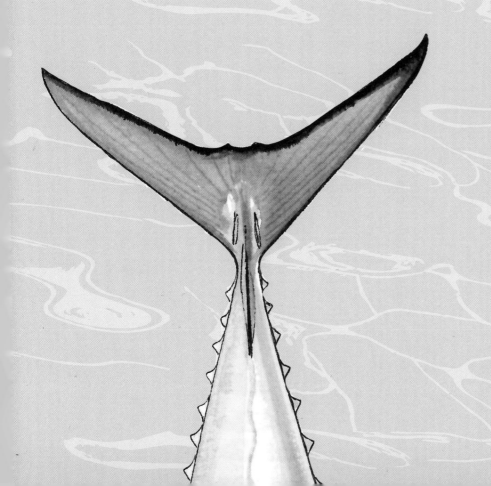

FISHES *of the* OPEN OCEAN

A NATURAL HISTORY *&* ILLUSTRATED GUIDE

FISHES *of the* OPEN OCEAN

A NATURAL HISTORY *&* ILLUSTRATED GUIDE

JULIAN PEPPERELL

ILLUSTRATED BY GUY HARVEY

THE UNIVERSITY OF CHICAGO PRESS
CHICAGO AND LONDON

The University of Chicago Press, Chicago 60637
University of New South Wales Press Ltd, Sydney, Australia
© Text: Julian Pepperell 2010
© Illustrations: Guy Harvey 2010
All rights reserved. Published 2010
Printed in China

19 18 17 16 15 14 13 12 11 10   1 2 3 4 5

ISBN-13: 978-0-226-65539-0 (cloth)

ISBN-10: 0-226-65539-3 (cloth)

Library of Congress Cataloging-in-Publication Data

Pepperell, Julian G.
    Fishes of the open ocean : a natural history and illustrated guide /
    Julian Pepperell ; illustrated by Guy Harvey.
        p. cm.
    Includes index.
    ISBN-13: 978-0-226-65539-0 (cloth : alk. paper)
    ISBN-10: 0-226-65539-3 (cloth : alk. paper) 1. Pelagic fishes.
    2. Deep-sea fishes. 3. Marine fishes. I. Harvey, Guy, 1955-
    II. Title.
    QL620.P46 2010
    597.177—dc22
                                        2009032290

∞ The paper used in this publication meets the minimum
requirements of the American National Standard for Information
Sciences—Permanence of Paper for Printed Library Materials, ANSI
Z39.48-1992.

JULIAN PEPPERELL, PhD, is one of Australia's best-known marine biologists and a recognized world authority on billfishes (marlin and sailfish), tuna and sharks. Julian has conducted research on these fishes in partnership with universities and governments for over thirty years and is an adjunct professor at a number of universities. He is a past president of the Australian Society for Fish Biology and recipient of the prestigious Conservation Award from the International Game Fish Association.

GUY HARVEY is a unique blend of artist, scientist, diver, angler and conservationist. In 1999 he collaborated with the Oceanographic Center of Nova Southeastern University to create the Guy Harvey Research Institute, providing scientific information for effective conservation and restoration of fish resources and biodiversity.

CONTENTS

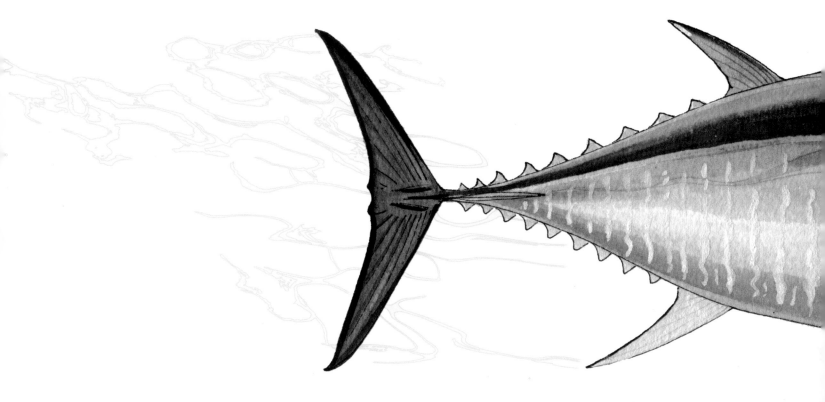

1 WHY FISHES OF THE OPEN OCEAN?

At last count, scientists had recognized more than 31 000 species of fish in the world, by far the largest order of vertebrate animals. However, of that number, only relatively few, perhaps several hundred species, might be thought of as fishes of the open ocean. Of course, this leads to the question of what exactly is the 'open ocean', and secondly, what defines the fishes that swim there.

Perhaps surprisingly, the open ocean has no widely accepted formal definition. In the context of this book, it is defined as the surface layer of the sea or ocean, to a depth of about 200 metres. The term 'open ocean' also conjures visions of the wide blue yonder, waters well away from land masses. In the context of this book, though, we will include in our definition of the open ocean waters over continental shelves, as well as waters around reef drop-offs, islands and atolls.

There have been a number of attempts to categorize marine fishes by habitat, all of which suffer from the same problem of fluid boundaries. Having said that, one system of classifying pelagic marine fishes, proposed by Russian ichthyologist Nikolai Parin, has considerable influence on the fishes selected for this book. Following Parin, I've grouped fishes into three categories according to the proportion of their lives spent in this open-ocean zone. This first grouping includes fish that spend their whole life cycle in the surface layer, or regularly move into this zone. The second group is fish that spend only a part of their life cycle in the open ocean. And those that temporarily visit the open ocean from other habitats comprise the third group. His three classifications, in a somewhat simplified form, are described below.

Permanent residents of the open ocean ('holoepipelagic' fishes)

Fish in this category include marlins, sailfish, tunas, dolphinfish, some flying fishes, sauries, some pomfrets, ocean sunfishes, the pelagic stingray, predatory pelagic sharks including the shortfin and longfin makos, the blue shark, silky and oceanic whitetip sharks and the surface filter-feeding sharks – the basking and whale sharks.

Below the habitat of these true surface dwellers live many midwater (mesopelagic) fishes, some of which make nightly excursions to surface waters and are therefore also included here. Many species in this category were hardly known at all before the advent of surface longlining – setting lines with hundreds or thousands of hooks suspended by buoys. This gear, set around the clock, not only catches the main target species (tuna, billfish, sharks) but also an assortment of these midwater

fishes. These include lancetfishes, oar-fishes, opahs, the louvar, and some of the snake mackerels (family Gempylidae) such as the oilfish, escolar and the snake mackerel itself.

Other types of fishes in this category strongly associate with surface materials, such as drifting weed, debris, logs or even drifting invertebrates such as jellyfish. Examples include the tripletail, pelagic leatherjackets and oceanic puffers. And finally, the pilotfish, a small member of the jack family, spends its life swimming in close proximity to large pelagic fishes, especially oceanic whitetip, blue and whale sharks. The remoras have adopted a similar habit, but, rather than swim near large hosts, they attach themselves by means of highly modified dorsal fins and are therefore passively transported with their hosts to cover the same geographic ranges.

Dwellers of the open ocean for only part of their life cycles ('meroepipelagic' fishes)

Fishes in this category might spend only their adult lives in this zone, or at least part of their adult lives. Examples would include Atlantic and chinook salmon, which enter the surface layer of the open ocean only as early adults, having migrated from fresh water. They feed and mature in the pelagic zone for a number of years before returning to freshwater rivers to spawn. Other species in this category included here are the whale shark, which is mostly oceanic but gives birth in shallow waters, and some flying fishes, which must attach eggs to floating algae such as kelp. Another group of fishes in this category spend only their juvenile phase in the surface layer. A good example is the bluebottle fish, which spends its juvenile life in close association with the floating colonial siphonophore, the bluebottle or Portuguese man-of-war, but when mature, takes up a benthic (bottom) existence. Two other major groups included here are some members of the jack family, Carangidae, and several of the Spanish mackerels (*Scomberomorus* spp.). Many jacks are found as juveniles around floating objects or buoys, while the juveniles of some, such as the golden trevally, mimic the pilotfish in looks and behavior, and in so doing, probably obtain similar protection and dispersal by swimming with their larger, highly mobile hosts.

Occasional visitors to the open ocean ('xenoepipelagic' fishes)

These are either coastal pelagic fishes which move offshore from time to time, or coastal species that associate with floating mats of weed or debris and drift into the open ocean as a result. Fishes in this category included here are some of the leatherjackets, chubs and some jacks, including, for example, the amberjack. Finally, also in this category are some species of fish which are normally coastal but which move offshore to take advantage of upwellings of nutrients or planktonic blooms, such as various herring and sardine species.

While these categories have helped guide the entries in this book, there remain some exceptions and outliers. A number of commercially or recreationally important species are largely coastal in their habitat preferences but may be encountered at the edge of the open-ocean habitat and are therefore included. In these cases, a general criterion for inclusion is that the species should be widespread, either occurring on both sides of ocean basins, or at least around offshore islands. Fishes in this category include the bluefish, bonefish, tarpon and roosterfish, a number of carangids (permit, queenfish, and some jacks/trevallies), most of the Spanish mackerels and several sharks and rays (bull shark, dusky shark and spotted eagle ray).

While relatively few fish species have adapted to life in the open ocean, those that have tend to be very successful. Many of the species included in this book have fully global distributions, occurring in all three major oceans as well as the Mediterranean Sea. The fishes of the open ocean depicted and described in these pages include the largest fishes on the planet – the whale shark, the basking shark and the manta ray. They also include the largest bony fishes in the world – the oceanic sunfish, the blue marlin, the black marlin and the Atlantic bluefin tuna. Some species, such as the streamlined wahoo and the sailfish, swim faster than any others, while several of the billfishes and tunas and sharks make regular transoceanic, or even interoceanic journeys. Some of these fishes, such as the skipjack tuna, are among the most prolific fish species on earth, supporting huge commercial fisheries that supply millions of people with affordable protein. The fishes of the open ocean are remarkable in many ways and it behoves us, as stewards of our marine resources, to manage and conserve them sustainably for the benefit and wonderment of all who will follow.

2 THE OCEANIC
WEB OF LIFE

There is an apocryphal story of an angler fishing one pleasant morning with high hopes of a great day's fishing. He hooks the first small baitfish of the day, but as he's winding it in, a bonito takes it. As the excited angler attempts to land the unexpected bonus, a big yellowtail promptly grabs the bonito. The sudden screaming of the little bait reel loudens even more when a large yellowfin tuna engulfs the yellowtail in one swoop and heads for the horizon. The ensuing fight is long and hard, and then, just when the tuna is nearly within range of the gaff, a huge blue marlin appears from nowhere, swallows the tuna and so the fight is on again. With incredible skill, and by this time, well out to sea, the angler finally has the catch of a lifetime alongside the boat when suddenly ... the little baitfish spits out the hook! Now, even if you don't believe this fish tale, it does illustrate the simple concept of a linear food chain quite nicely. Old textbooks on biology often referred to these as food chains, in which each link connected different organisms in a direct line. But in reality, the way that organisms rely on each other for food is far more complex. Hence the term 'food web' was coined, quite aptly relaying the idea of an intricate interdependence of a whole suite of plants and animals within a vast, but relatively closed ecosystem. Here, we will explore the food web of the focus of this book – the food web of the upper layers of the open ocean.

Ecosystems of pelagic fishes

Pelagic fishes live in a somewhat unusual environment, often far from land, with little if any physical structure available. This habitat has been likened to a marine desert, since the availability of food is often very patchy, dependent on transport of nutrients from the land or in upwellings from the abyssal depths. Some areas may be rich in nutrients, and therefore are hotspots for marine life, while others may not contain much life at all. This world in which open-ocean fishes live is a complex self-contained ecosystem that maintains a rich and diverse fauna and flora on a vast scale. This environment extends from the ocean's surface to perhaps 200 metres below – in reality a mere film on top of the world's great oceans. But this thin and delicate layer contains the stuff of life – sunlight, oxygen and warmth.

The building blocks of all oceanic food webs are the phytoplankton – tiny single- or multi-celled floating plants which manufacture sugars by photosyn-

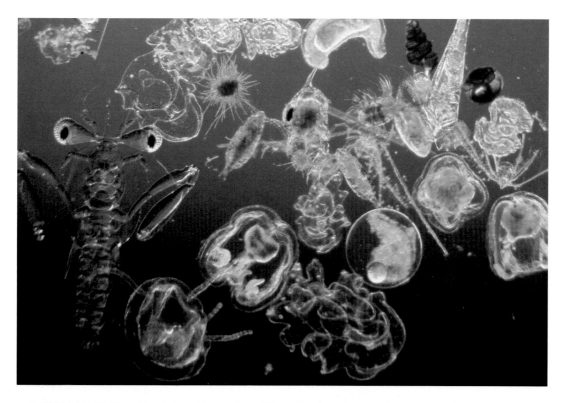

Top Tiny phytoplankton and zooplankton are key building blocks in the oceanic food web.
seapics.com/ iq3-d/Peter Parks

Middle A feeding frenzy of small tunas sends signals to other predators such as sharks, dolphins and seabirds.
Julian Pepperell

Bottom A searching gannet hunts for surface-feeding tuna.
Julian Pepperell

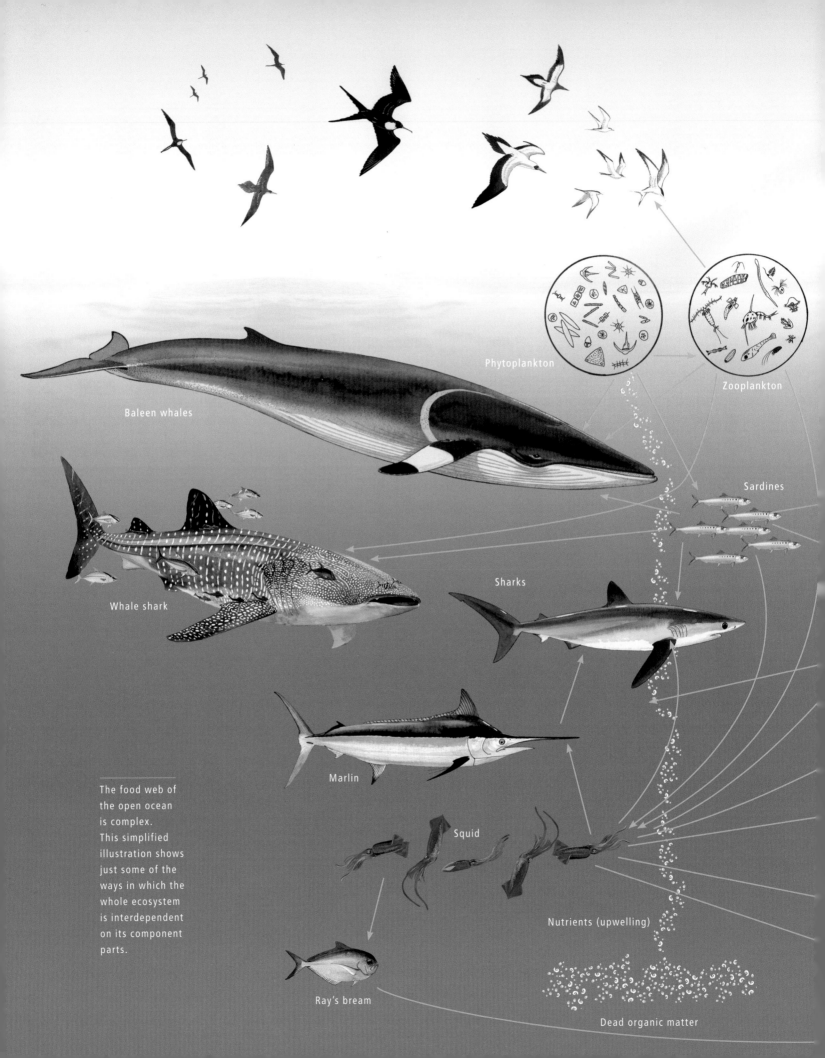

Phytoplankton

Zooplankton

Baleen whales

Sardines

Whale shark

Sharks

Marlin

Squid

The food web of the open ocean is complex. This simplified illustration shows just some of the ways in which the whole ecosystem is interdependent on its component parts.

Nutrients (upwelling)

Ray's bream

Dead organic matter

Jellyfish

Sardines

Turtle

Dolphins

Marlin

Sharks

Scads

Tuna

Sunfish

Sperm whales

Lancetfish

Swordfish

thesis. Nutrients are also necessary for plant growth, which in the open ocean are mainly derived from organic matter (dead animals and plants) formed by decomposition on the ocean floor and brought to the upper layers by upwelling currents. The abundant phytoplankton form the vast pastures which are grazed upon by the zooplankton – tiny animals including the larval forms of many invertebrates, especially crustaceans, and fishes. As well, the phytoplankton are the primary source of food for the teeming schools of filter-feeding anchovies, herrings and pilchards, not to mention probably the most abundant animal on earth, krill.

Oceanic food webs can be rather complicated. Small baitfishes such as anchovies, pilchards and sauries are the staple diet of a whole array of larger animals, including many species of predatory fish. These include tuna, sharks, dolphinfish, mackerel, wahoo, marlin, sailfish and swordfish.

From the time they are only a few centimetres long, all of these species hunt the seemingly defenceless baitfish schools. Other predators also take their turn to feast on this food supply. Dolphins and porpoises, which collectively number in their tens of millions, take their considerable share, as do many species of seabird, each specialized in the size and type of food they target. In fact, some species of seabird only eat small baitfish which are driven to the surface by feeding tuna. Without the assistance of tuna, the birds would have no source of food.

Squid also feast on anchovies and small fish, but they are also huge consumers of zooplankton, particularly the teeming masses of tiny crustaceans which form dense concentrations in the water column (clearly visible at night on any good depth sounder as this so-called scattering layer rises to the surface). And of course, the many species of oceanic squid themselves are a vital, often dominant component of the diets of many animals in different levels of the food web. Sperm whales, for example, eat squid almost exclusively, while broadbill swordfish are also particularly dependent on this abundant supply of protein.

Other components of this tangled web include marine turtles, which eat not only small fish, but also jellyfish and jellyfish-like plankton such as salps. Then there are the mighty baleen whales which filter great gulps of water through their

The largest animals of the open ocean are all filter feeders. Here, humpback whales lunge upwards through schools of anchovies, taking huge mouthfuls of water which they strain through their baleen plates.
seapics.com/ James Watt

baleen, or 'whalebone' plates. In the Arctic and Antarctic, these whales graze easily on krill, while in more temperate waters they take advantage of concentrations of baitfishes. Witness the annual appearance of dense schools of pilchards off the South African and Western Australian coasts which are always attended by pods of hungry minke whales.

On the subject of filter feeders, the largest fishes on the planet, the basking and whale sharks, also feed on the smallest organisms. Basking sharks do so by simply swimming through the plankton with mouth agape, while whale sharks more actively suck in mouthfuls of water containing small crustaceans and schooling fish, sometimes engulfing larger predators such as tuna in the process.

As mentioned, sharks are important components of the food web, consuming small fishes and squid in large quantities. But larger predatory sharks also prey on animals further up the web. As adults, some, such as the white shark, become specialized feeders on marine mammals, including seals, dolphins and even whales. The tiger shark has evolved the singular ability to bite through marine turtle shells – a particular favorite food item of this widespread shark. Even at the apex of one part of the food web, the large marlins and tunas are targeted food items for other predators, including large shortfin mako sharks and one particular species of toothed whale, the false killer whale. The latter appears to be a specialist predator of large oceanic fishes, having

Marine turtles wander over vast distances of open ocean consuming small fish and jellyfish along the way.
marinethemes.com/ David Fleetham

Jellyfishes
and other
coelenterates
are a significant
proportion of
macroplanktonic
biomass and
are eaten by a
surprising variety
of fishes and other
marine animals.
seapics.com/
Masa Ushioda

a very wide gape and relatively large teeth compared with other toothed whales.

Swimming the gauntlet

One fascinating aspect of the oceanic food web which makes it very different from land-based food webs is the general principle of larger fish eating smaller fish regardless of species. Many land animals, such as mammals and birds, produce large, well-developed young, and care for them until they are virtually adult-sized and have a good chance of survival. In the ocean, this is rarely the case. Most marine fish produce millions of eggs each year, which are simply released into the water, fertilized and left at the mercies of the currents. As each tiny larval fish grows towards adulthood, it must run a perilous gauntlet. Baby tuna and marlin can and will be eaten by any larger predatory fish, including larger tuna and marlin. It is not uncommon to find baby billfish inside the stomachs of species like dolphinfish, while small marlin bills can often be found embedded in the stomach lining of adult marlin. The danger of being small is probably why so many pelagic fishes have very rapid growth rates. 'Grow or die' could well be the motto of the mixed layer. It is indeed a fish-eat-fish world.

Those who fish can gain some idea of the food web immediately influencing their catch by conducting simple dietary studies on landed fish. Open the stomach of any pelagic fish and chances are you will find a fairly broad range of food items. The main food groups you might expect to see would be crustaceans, molluscs and, of course, fish. The crustaceans might include the larvae of crabs (a very common dietary item of tuna and, sometimes, marlin), free-swimming crabs, prawns, shrimps and, quite commonly, stomatopods – praying mantis-like creatures also known as prawn killers. Molluscs in the pelagic fish diet consist almost entirely of cephalopods; that is, squid, cuttlefish, octopuses (including paper nautilus and many species of pelagic octopus) and the occasional tiger nautilus. Often, the indigestible beaks of squid are all that remain of this very important group of prey. Most stomachs will contain fish, or at least the remains of fish, sometimes, one species of fish completely dominating the diet. I well remember the first sailfish I dissected, the stomach of which was completely crammed with small pelagic toadfish. On other occasions, several different species of fish of different sizes might be found. For example, large blue marlin in Hawaii are often found to contain very small boxfish and triggerfish in their stomachs. Since marlin do not possess comb-like gill rakers which other species use to sift small food items from the water column, it would appear that large marlin pick off these tiny fishes, one by one.

Breaking the web

Marine ecosystems were in existence for hundreds of millions of years before human predators began to take ever-increasing quantities of organisms from different levels of the food web. Looking at just some of the fisheries of the open ocean which have developed only in the past few decades, annual catches of several levels of food web animals have reached staggering levels. Global annual catches of skipjack tuna have recently topped two million tonnes, over half

One of millions. The developing egg of a dolphinfish, Coryphaena hippurus. *Many open-ocean fishes produce huge numbers of eggs that float near the surface, where they are fed upon by other fishes and specialist seabirds. Syd Kraul, Pacific Planktonics*

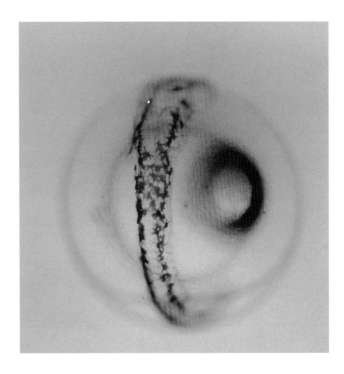

of which is taken in the western Pacific Ocean. More than one million tonnes of yellowfin tuna are also removed annually from the world's oceans, a high proportion of the catch consisting of juvenile fish. Catches of sardines and squid would each also top the million tonne mark, while the take of smaller fishes such as anchovies would be well over 10 million tonnes.

The Peruvian anchovy holds the dubious distinction of being the most heavily exploited fish. The fishery for this tiny filter feeder, which was mainly converted into dried fishmeal, increased to a peak of over 13 million tonnes in 1971, but then almost completely collapsed and catches remained very low until the 1990s. The primary cause of the collapse is now thought to be a strong El Niño event, but many also maintain that intensive fishing must take at least some of the blame. Regardless of what or who caused the demise of these baitfish, it was clear that the animals which depended on the baitfish, such as seabirds and larger fishes, were affected when their primary food source diminished so dramatically.

The removal of such huge quantities of biomass from oceanic food webs raises two very big questions: Are such catches sustainable, and do they have any effect on the food webs themselves? For example, is there any discernible effect on prey species of removing large numbers of predatory fishes such as tuna and billfish, or conversely, does the removal of many thousands of tonnes of baitfish, squid or krill have any effect on predators higher up the web? Other secondary questions might include: Are these prey items, in terms of both species and size, being actively selected, or is the feeding of billfishes and other predators primarily (or entirely) opportunistic? Is prey (food) limiting at any stage of the life cycle of predatory fishes? Do populations of predatory fishes 'boom or bust' depending on prey availability? What is the total biomass of predatory fishes in a given area at a given time, and what biomass of prey is required to sustain the predator biomass?

Models of the oceanic ecosystem

Historically, fisheries science has tended to be based on single species. For example, studies on striped bass, summer flounder or skipjack tuna would concentrate on considerations of the life cycle of that particular species – its reproduction, age and growth rates, and mortality rates (from natural causes and from fishing) being the primary concerns. Dietary studies might be done also, but these would be secondary to the central study. Such

Food items taken from the stomach of a juvenile longtail tuna, *Thunnus tonggol*. Note the similar sizes of the organisms, including fish, crustaceans and squid, suggesting that the tuna has used its gill rakers to strain these items from the water column.
Julian Pepperell

A feeding school
of skipjack tuna,
probably the most
prolific small
predator of the
open ocean
Julian Pepperell

an approach would result in a reasonable understanding of the biology of a particular species, but in isolation from the rest of its ecosystem. Studies might even have included the effects of environmental variables on the organism, such as temperature or currents, but effects of or on other organisms were rarely considered in detail.

As our concept of ecosystems has improved, so too has the way in which animals within an ecosystem are studied. And with the great advances in computing power over the past several decades, sophisticated 'models' of entire ecosystems, and therefore food webs, can now be generated.

Such models can be used, at least in theory, to see what might happen if fish or other organisms are depleted at a given trophic level. However, a major problem with such computer models is that they rely completely on the quality of the information fed into them. So, while we have fairly good information on the biology and life cycles of some of the more commercially important species, such as skipjack and yellowfin tuna, little is known about critical parts of the food web, such as the biology of zooplankton, or of the

many species of oceanic squid. As the saying goes, 'garbage in, garbage out', so with such major uncertainties about key elements, these models as they stand may be interesting, but they are still some way off from being used as predictive tools in fisheries management. What the models do show is that we need to know a lot more about the basic biology of many of the building blocks of food webs. In fact, so little is known about the biology of many of the building blocks of the oceanic food web that caution must be the primary concern in any development of any new fishery within this complex system.

Competing with other predators

One interesting argument has arisen from time to time to suggest that exploitation of fish and other organisms by humans is not necessarily a major problem. This stems from the idea that all of the animals which eat fish consume such huge quantities that the human part of the equation must be relatively minor. Such comparisons suggest that quantities consumed by seals, whales, dolphins and seabirds may even be one or more orders of magnitude greater than commercial (and recreational) catches.

While these arguments imply direct competition between humans and marine animals for the same resources, several important points need to be made. Commercial and recreational catches consist largely of finfish, while the diets of whales, dolphins and seabirds include high proportions of cephalopods (squid and octopus) and crustaceans. Seals certainly target fish, some of which are also caught by people, but they also eat a lot of squid. As well, many of the fishes being consumed by marine mammals and birds are not the same species as those targeted by humans, or if they are, they are not caught at the same sizes. Many of the most abundant species of seabirds, for example, eat very small fishes, or even fish eggs, while it is also known that whales rarely compete directly with humans for the same fish stocks.

So, even though these sorts of simple comparisons don't tell us a lot about direct competition between humans, fishes and other marine animals, they do help to focus on the connections between human use and the oceanic food web. We know that predation at all levels within the food web has existed in the oceans for eons, and that such predation is the cause of a large component of what is called 'natural mortality' of fishes and other marine organisms. The very recent introduction of the human element into ancient food webs is a new and poorly understood additional pressure on these natural systems and it is therefore risky to increase exploitation on already existing large scales before we have a fuller understanding of the whole picture. We can coexist with and within these food webs. We just have to learn more, and take care at every step along the way.

3 FORM AND FUNCTION

As outlined in chapter 1, many different kinds of fish have adapted to life in the surface layers of the open ocean. The term 'pelagic fishes' covers a broad range of animals, from sardines to sunfish to swordfish, so this chapter will focus on the forms and functions of three of the most successful, highly specialized groups of large, predatory pelagic fishes – the billfishes, tunas and lamnid, or mackerel sharks. It will examine the range of adaptations that these fishes have acquired and consider how a number of these remarkable features, so well suited for life in the open ocean, have evolved independently in these largely unrelated pelagic fish groups.

The process of evolution is often popularly explained as 'the survival of the fittest', a phrase which certainly helps to explain how a similar body shape would arise in a number of unrelated animals, including fishes. Think about the body shapes of many marine animals which live in the surface layers of the ocean and which have predatory lifestyles. Dolphins and porpoises, in particular, have very similar body shapes and forms to those of swordfish, marlin, tuna and mako sharks. However, dolphins are not fish at all, but mammals which evolved from four-legged terrestrial ancestors while still retaining telltale characteristic features such as mammary glands (they suckle their young) and hair (albeit, just a few sensory hairs on their snout). And to emphasize this phenomenon even further, a group of extinct marine reptiles, the ichthyosaurs, which also evolved from four-legged reptilian ancestors, were extremely similar to dolphins, sharks and billfish in body shape. These fabulous creatures dominated the oceans, occupying a virtually identical niche to dolphins and killer whales of today. It is now clear to scientists that all of these animals – swordfish, marlin, tuna, sharks, dolphins and ichthyosaurs – evolved from quite different origins but all came to resemble each other, at least superficially, because this particular body plan works best for the open-water environment, and evolution has therefore selected this shape many times over the ages.

The development of this kind of similarity in body shape and function in unrelated or distantly related animals is known as parallel or convergent evolution, and the mako shark and broadbill swordfish provide a particularly good example of this phenomenon. Looking at the general body plan of broadbill swordfish and mako sharks, one can see many similarities. Both have very similar fin structure and placement, tough, scaleless skin, very broad, single lateral keels on the caudal peduncle or tail wrist and a large, symmetrical tail. However, having noted such remarkable similarities, closer examination of the anatomy of both

No doubt ichthyologists would relegate him [the mako] to the shark family, and I was compelled to do that also, but I never saw a shark before with any of this one's marked features. He actually had something of the look of a broadbill swordfish without the sword ... Here was a sea creature, an engine of destruction, developed to the nth degree. I had never seen its like. Even an Orca [killer whale] could not do any more ravaging among sea fish.

The body shapes of fishes tell us much about their lifestyles. A flounder is obviously built for life on the bottom, lying as it does on its side, with both eyes on the same side of the head keeping watch from a camouflaged position. Fishes with very large paddle-shaped tails and pectoral fins are often found around reefs in relatively shallow water. They use these broad paddles to move effortlessly around their environment, holding their own against surging tides and currents. And so too, in the surface layers of the open ocean, predatory pelagic fishes have evolved and adapted superbly to their mode of existence, evolving sleek, streamlined bodies for life in the fast lane.

The open-water habitat presents some unique problems for fishes. Availability of food across the vast oceanic distances is very patchy, and in fact, the surface waters of major oceans are often termed 'marine deserts'. This is because nutrients sink to the bottom, and only become available to surface dwellers in widely separated regions of upwelling. Pelagic fishes therefore need to be able to travel long distances to find food, and predatory species such as tunas, billfishes and sharks also need to be able to move very rapidly when needed, as their main food items, smaller fish, also move quickly.

Compared with air, water is a dense medium which is difficult to move through. In fact, water is 800 times more dense than air, creating obstacles to rapid movement. Yet we know that pelagic fishes do move through this dense liquid

The similar body shapes, including fin placement, tail shape and broad caudal keels, of the shortfin mako shark and the broadbill swordfish. A perfect illustration of convergent evolution in the open ocean.
(Top) *Chris and Monique Fallows/ Oceanwideimages.com*
(Bottom) *marinethemes.com/ Steve Drogin*

reveals that broadbill swordfish are bony or ray-finned fishes (teleosts) whereas all sharks and rays, including makos, are elasmobranchs, and have mainly cartilaginous skeletons. The relationship between swordfish and mako sharks is therefore distant, and the common ancestor of both would not have resembled the streamlined form of either. In 1925, famous writer and game fishing pioneer Zane Gray, in *Tales of the Angler's Eldorado*, described the capture of his first mako shark, weighing 117 kg (258 lb), off the North Island of New Zealand. Struck by its similarity to the swordfish, he was inspired to write:

with relative ease, often at speeds which are hard to comprehend. Although very few studies have actually been carried out on swimming speeds of open-ocean fishes, those that have suggest that species such as sailfish and wahoo may attain speeds as high as 100 km per hour in short bursts. We also know from tagging programs that many species of tuna, billfishes and sharks are capable of traveling vast distances, albeit at much slower speeds. Let's look at some of the features which enable these remarkable fishes to perform such feats.

Consider firstly the general body coloration of pelagic fishes. Virtually all species which spend most of their time at the surface, especially during daylight hours, are heavily countershaded; that is, they are invariably very dark on the back and pale or white on the lower flanks and belly. One widely held view for many years was that this countershading was for camouflage when a fish is viewed either from above or below. The theory behind this is that the dark backs of fish would be more difficult to see when viewed against the bottom, while the pale or white belly would merge with the lighter background of the sky when viewed from below. This is a reasonable explanation of the two-toned coloration of bottom-dwelling species, but it is not a good reason for surface-dwelling or pelagic fishes to be so markedly two-toned.

In open water, there is nowhere to hide, so for both predator and prey, it would be a decided advantage to be hard to see. And in fact, experiments have shown that this is exactly what countershading does. It makes a fish much harder to see when viewed not from the top or bottom, but from the side. In shallow or surface waters, sunlight filters down from above, creating a shaded undersurface to a solid, roughly cylindrical or ovoid object such as a fish, dolphin or penguin. The shaded undersurface merges with the often mottled sides and darker back, resulting in the object appearing to be a uniform shade of gray, or even light blue. This phenomenon is readily observable when snorkeling or diving. When viewed from the side, surface-swimming fishes

such as halfbeaks, barracuda and needlefish almost disappear, blending eerily with the filtered, blue light. Using this countershading to become almost ghost-like, many of the predatory pelagic fishes then employ a further tactic to confuse their intended prey. They produce either very dark, or highly reflective, temporary vertical stripes or bars on their bodies which serve to break up any residual outline of the body.

In addition to countershading and vertical barring, a number of surface fishes have evolved very shiny, almost metallic, flat bodies which reflect light

Top Many pelagic species of fish show marked counter-shading coloration. Here, Atlantic bluefin tuna show dark blue dorsal surfaces contrasting with white flanks and bellies.
Boyceimages.com

Bottom A striped marlin showing both counter-shading and vertical barring, both common attributes of many species of pelagic fishes.
Guy Harvey

very efficiently. Experiments have shown that a mirror suspended in the water reflects light of identical color and intensity to the surrounds, causing it to appear transparent. The dolphinfish, with its highly compressed body, does just this, and even though it is brightly colored, when viewed from some angles underwater, it takes on a silvery, highly reflective appearance from other angles. But the real experts at mirror camouflage are baitfishes such as herring and sardines. These can perfectly reflect surrounding laterally directed light, thereby almost disappearing from view in a piscatorial 'smoke and mirrors' magic trick.

Now consider the body shapes of pelagic fishes. We have already noted some similarities such as fin placement, and while there is a fair amount of variability among species, there is a strongly recurrent theme of pointed snouts (developed to the extreme in the billfishes), rounded bodies (in cross-section) and deeply forked, or 'lunate' tails. The technical term for this body shape is 'fusiform', although it is also often called 'thunniform', which simply means tuna-shaped. In essence, this classic body shape of many fast-swimming pelagic fishes is optimized to reduce drag. Mathematically, this is achieved if the widest part of the body is about two-fifths of the distance from the snout to the tail, or put another way, if the width to length ratio of the body is about 0.25. Some call this a 'teardrop' shape, but perhaps torpedo-shaped is a better descriptor. In any case, fishes which fit this body plan would certainly include all of the tunas, the lamnid sharks (mako, white, porbeagle and salmon shark) and the billfishes (with the possible exception of the long, slender spearfishes).

While many fishes swim by means of body undulations, thunniform fishes move through the water by keeping their bodies relatively stiff and transferring nearly all of their considerable muscular power to the tail. And under high speed, their tails beat very rapidly indeed. Indicating a common function, the tails of many (but not all) pelagic fishes are very similar in shape. Tails of large tunas,

billfish and mackerel sharks (as well as dolphinfish, trevallies, and so on) are all relatively large, swept back, and narrow-bladed, a classic shape termed 'lunate'. Interestingly, the tails of pelagic fishes have become more slender with evolution, with the relatively 'primitive' mackerels (Spanish, king, cero, and so on) having quite broad lobes, while those of the highly advanced tunas are more narrow.

The tail is certainly the propeller which drives the fish forward, but other parts of the anatomy in the tail region also help to achieve this by reducing drag. Lateral keels are very prevalent among pelagic fishes, on either the tail wrist (the caudal peduncle) or the tail itself. All of the marlins, sailfish and spearfish have paired keels on either side of the tail wrist, while, as noted above, broadbill swordfish and mako sharks have a single very wide keel on each side. Tunas also have a prominent single keel on the tail wrist, and usually two smaller keels at

The characteristic finlets of tunas – in this case, a yellowfin tuna – and other pelagic fishes are thought to reduce turbulence and drag caused by the main dorsal and anal fins as they cut through the water.
Julian Pepperell

the base of the tail itself. In all cases, the function of these keels is undoubtedly to slice through the water as the tail swings rapidly from side to side, effectively reducing resistance from the water and providing a much more efficient tail beat.

Another common feature of fast-swimming pelagic fishes is their possession of so-called finlets, or much smaller dorsal and anal fins towards the tail. Take yellowfin tuna, for example. This species has eight or nine bright yellow finlets leading towards the tail on both the upper and lower surfaces. These move from side to side in unison, and while their function has never been fully explained, primarily because no one has taken the time to carry out the definitive experiments and observations, their presence almost certainly reduces turbulence, and therefore drag, by breaking up eddies of water formed by the larger dorsal and anal fins in front of the finlets as the fish moves forward.

While all of the tunas (as well as trevallies, mackerels and wahoo) have multiple finlets, billfish, including broadbill swordfish and lamnid sharks, only have single 'finlets' which are simply very small second dorsal fins. Again, how a single finlet could help reduce drag is not entirely clear, but it is obviously there for a very good reason since it has evolved in three such separate, unrelated fishes.

Another related feature of two separate groups of pelagic fishes, the istiophorid billfishes and the tunas, is the way that their fins fold down so neatly and perfectly into slots and depressions. The first (main) dorsal fins and the pelvic fins of all of the tunas and the billfishes (with the notable exception of the broadbill swordfish) fold flush into slots or depressions, while the istiophorid billfishes have gone one step further with beautifully foldable first anal fins as well. Like the classic F111 aircraft, these fins fold away for ultra-high-speed swimming, with the fish at the same time holding their pectoral fins against the flanks. In the case of tunas, the pectoral fins fold neatly into cowling-like depressions, whereas the billfishes have not developed this particular feature.

As usual, there is an exception to the rule. The black marlin has completely rigid pectoral fins set in place like two swept-back airfoils, providing lift to the particularly large head and shoulders of this species. When small, black marlin can and do fold their pectoral fins against the body, but from a size of about 15 kg (33 lb), the pectoral joint calcifies, permanently locking the fins in place. Broadbill swordfish have very similar fin arrangements to marlin, but with no mechanism at all for folding any of their fins. In this sense, broadbill are much more akin to the lamnid sharks, with the similarity in shape and form with makos already noted above.

So far, we have been looking at external characteristics of pelagic fishes, but they also possess many internal features which again show extreme adaptations to the requirements of living in such a demanding environment. For example, many pelagic fishes have relatively large amounts of red muscle compared with white muscle, which tends to extend into the interior of the body towards the spine. While white muscle is utilized for burst activity, red muscle is used for sustained swimming, so species with lots

The sleek shape of a pair of yellowfin tuna with their first dorsal fins folded completely into slots for high-speed swimming
Boyceimages.com

of these fibres, especially if internalized, are likely to be long-distance swimmers. Indeed, species known to travel over thousands of miles, such as skipjack tuna, the bluefin tunas and mako sharks, all have relatively large amounts of red muscle, readily seen in a cross-section of the body.

Swimming fast requires a lot of energy and, therefore, a lot of oxygen. Once again, we find that pelagic fishes have evolved highly refined mechanisms for extracting oxygen from the water. Firstly, they tend to be ram ventilators, meaning that they must keep swimming with mouths agape in order for water to pass rapidly over their gills. Moving gill covers back and forth to suck in water uses considerable muscular energy, so ram ventilation has developed in different pelagic species to conserve energy reserves. The gills themselves have very large surface areas in most pelagic species (again, convergently evolved), allowing them to extract more oxygen from passing water than other fishes. And finally, the blood of tunas in particular is highly specialized, containing two to three times the hemoglobin content of most inshore fishes. This not only enhances their high metabolic lifestyles, but also enables species like bigeye tuna to dive into depths where oxygen concentrations are extremely low and still function efficiently.

If these features weren't enough, one of the most interesting aspects of the evolution of pelagic fish is that many of them have evolved the ability to retain heat within their body cores (pelagic fishes being the only fish to have done so). This is achieved by a countercurrent blood circulatory system which positions major blood vessels alongside each other. Heat is then transferred from blood leading away from the warm body core (warmed from muscular and digestive activities) to blood coming from the colder gills. Such warm-bloodedness has evolved to the highest degree in the advanced tunas, such as the bluefins and the bigeye, and also in the lamnid sharks in particular. Marlins have not developed this ability to a great extent, probably since they tend to remain in warm water for most of their lives.

Controlling body temperature has many advantages, but controlling the temperature of the brain and the eye has even more. And so it is that a remarkable brain/eye heater organ has arisen independently in several groups of pelagic fishes. The brain/eye heater was first discovered in swordfish, and it took some time to realize what it was. The organ itself is actually a modified muscle which was once used to control movements of the eyeball. However, through evolutionary timescales, the role of the particular muscle was taken over by other muscles, allowing it to gradually lose all of its ability to contract, and to instead become a heat-generating mass of tissue behind the eyeball. In this position, the modified muscle can warm the back of the eye (and hence, the retina) and also the brain of the fish to temperatures up to 15°C higher than surrounding water. This keeps both the eye and the brain working efficiently at cold temperatures, especially when the fish dives deep. Prey fish do not possess such a brain heater, and so it is thought that swordfish gain a distinct advantage of vision and reaction time over their prey. Other pelagic fishes, including tunas, marlin and opah, have also developed brain heaters, but to a lesser extent. The exception to this rule, however, is the primitive tuna-like fish, the butterfly mackerel – a species which lives in subantarctic waters. This fish has also developed a particularly large brain/eye heater from entirely different muscles than those of the swordfish.

The actual functions of many of the features of pelagic fishes outlined above are often speculated about, but with not a lot of empirical evidence to prove one way or another. However, one recent scientific study considered some of these questions in detail, showing that the similarities of body form in at least two very different types of pelagic fish are more than just skin deep. In a 2004 study on mechanical design of lamnid sharks and tunas, researchers at the Scripps Institution of Oceanography monitored the muscles of mako sharks swimming in a tank under

Striped marlin
showing first
dorsal fin partially
folded for rapid
swimming (top)
and with all fins
fully erect for
rapid direction
changes (bottom)
Boyceimages.com

controlled conditions. They also carefully dissected the muscles and tendons of makos and found some remarkable similarities with another species, the bluefin tuna. They found that mako sharks have essentially the same swimming motion as bluefin, with the most side-to-side motion in the tail region while the body is held relatively rigid. They also found that arrangement of muscles in both species was remarkably similar, with very long 'cones' of muscle fibres leading down to tendons in the narrow caudal peduncle. These adaptations were shown to be unlike virtually any other fishes (although presumably the billfishes would probably also show similar features), yet again demonstrating the remarkable phenomenon of convergent evolution in unrelated pelagic fishes.

This chapter can only briefly outline some of the adaptations and features of just a few of the open-ocean fishes. The study of the physiology of these animals has been hindered in the past because of the difficulties of keeping them in captivity under controlled conditions. The last few years, however, have witnessed a revolution in the way we can study these fishes, brought about largely by the development of archival and popup satellite tags. Hundreds of tuna and billfishes are now being tagged with these mini-computers, and the results, which tell not only of their migrations but also of their amazing abilities to dive to depths never before thought possible, are rewriting what we know about them on a regular basis. Many of the results of these studies so far are outlined in the species treatments which follow.

4 INTERACTIONS
WITH HUMANS

As we have seen, many of the fishes of the open ocean dwell in the surface waters of the world where their environment is warm, suffused with oxygen, lit by filtered sunlight and full of life. Many of these species prefer to stay within this friendly, photic zone by means of daily and seasonal migratory behavior. Few of them associate for very long with land masses or the ocean floor, and many are true nomadic wanderers, sometimes called 'fish without a country'. This was recognized as long ago as 1757, when Fray Martin Sarmiento wrote: 'Los atunes no tienen patria, ni domicilio constante. Todo el mar es patria ellos. Son unos peces errantes', which means 'Tuna have no native country, nor lasting home. All the sea is their native country. They are wandering fish'. This concept has persisted to the extent that, today, many of the tunas, billfishes and other oceanic fish are classified as 'highly migratory' under the United Nations Convention on the Law of the Sea (UNCLOS).

Although many of these fishes spend much of their life cycles beyond the continental shelves of the great land masses, some interactions with humans have been recorded from very early times. Depictions of swordfish (*Xiphias gladius*) appear on Egyptian friezes, while the Romans fished for giant bluefin tuna (*Thunnus thynnus*) and swordfish in the Mediterranean using methods which

are still in use today. Swordfish were hunted on the surface by harpoon, from vessels painted and shaped to resemble their quarry. Pliny describes these vessels and explains that they were designed to enrage the swordfish so that they would come to the surface and charge the boats, whereupon they could be harpooned.

The regular migratory routes of bluefin tuna around the Mediterranean were so predictable that each year for probably thousands of years, huge set nets were staked out to trap some of the passing schools. These are the famous trap fisheries for giant bluefin tuna of the Mediterranean which go under different names in different countries. In Italy the fishery is called the 'tonnara', in France, Tunisia and Algeria, the 'madrague' and in Spain, the 'almadraba'. In this unique fishery, arrays of large nets are anchored off many strategic points of land throughout the Mediterranean during the known traveling season of the giant bluefin. The nets entrap schools of tuna, which are then encircled by wooden boats and free-gaffed as the nets are hauled in.

Fortunately, some areas where nets have been set have maintained almost continuous records for over four hundred years while many others go back at least two centuries. These records are particularly useful in determining abundance of tuna through time since the fishery is so widespread geographically. This means

An early 19th century German illustration of the annual ritual bluefin tuna fishing activity known as the tonnara. This fishery has been in continuous operation for thousands of years.

that if similar fluctuations in catches have occurred at different trapping points around the Mediterranean there is a strong chance that such fluctuations really do reflect the abundance of fish at that time and not merely the chance of catching fish. We are always wondering what fish populations might have been like before the advent of intensive fishing. This unique set of records of a tuna fishery spanning four centuries might provide a key to unlock some of those mysteries.

Recent sophisticated analyses by French scientists using the latest powerful statistical and computer techniques have provided insight into such critical questions. These show that the catches of bluefin tuna in the Mediterranean have fluctuated enormously throughout this long period. In some years, catches have been up to seven times those of other years in a period pre-dating other fishing for Atlantic bluefin tuna. Another fascinating pattern emerged, indicating what appear to be true cycles of abundance – with one cycle being about 100 years long and another repeating about

every 20 years. Results indicate that there is indeed a strong correlation between catches and actual abundance of tunas, that non-yearly spawning can explain a lot of this variability and that external influences, such as variations in solar irradiance, may well have very strong effects on tuna numbers. This is not to say, of course, that fishing has not had a large impact on bluefin tuna numbers in recent decades, but it does suggest that fish numbers would vary enormously even in the absence of human activity.

Pacific, Atlantic and Indian Ocean island peoples have long had a special relationship with open-ocean fishes, including tuna, billfish and sharks. Island dwellers depend on fish for protein, and the oceanic fishes, in turn, are attracted to islands. And because these fishes are constantly on the move, the supply of fish was continually replenished. Pacific islanders developed specialized methods for fishing for tuna, dolphinfish, wahoo and even marlin. Specialized canoes and equipment were designed, and beautiful lures fashioned from opalescent shell, bone and feathers. In Hawaii, catching a large tuna was considered a demonstration of prowess, while in the Solomon Islands large sharks are revered as spirits of ancestors. There is good evidence of people in the Society and Marquesas Islands catching small tuna (probably skipjack tuna and frigate mackerel) and of dolphinfish being commonly caught in some areas. There are also stories of islanders catching billfishes, and even whales, but there is no physical evidence of these encounters. Except, that is, in one place. Apparent systematic catches of marlin, over a long period of time, did occur in the Marianas Islands (part of Micronesia), on the island of Rota in particular. Marlin bones occur in middens there dating right through the archaeological record from about 3000 years ago until the Spanish colonial period. The marlin bones were not identified to species level, but since blue marlin are the dominant species in this area, it is reasonable to assume that these ancient fishermen were enjoying the thrill of catching blue marlin all those years ago.

One fascinating legend from Papua New Guinea involves tuna and the village of Pari (near modern-day Port Moresby). Near this village, it is said that a woman gave birth to five tuna at the shore of a semi-enclosed bay, now called Oyster Bay, which opens off another Bay now called Bootless Inlet. Concealing this unusual event from her husband, she released the tuna into the bay and arranged to suckle them every day, summoning them by breaking a mangrove twig. Her husband discovered the fish by accidentally breaking a twig by the shore himself, and when they came towards the sound, he speared one and brought it home for a meal, which his anguished wife refused to eat. She subsequently sent the remaining four tuna away to sea for safety and later she revealed to her husband that she had borne the tuna, and that he had killed his own child. As a result, tuna fishing, which is highly seasonal, was both preceded and accompanied by strict taboos and many ritual activities. These included separation and sexual abstinence of intending fishers and rituals accompanying every

activity from the collection of materials for making nets through to the cooking of the fish (tuna were not allowed to be speared but had to be taken from the water by hand). Moral behavior in the village was said to affect the success of the fishing expeditions. The identity of Pari continues to be linked to tuna fishing in particular, and to the phenomenon of an annual journey by the fish from the sea into the bay where they swim an anti-clockwise circuit to be corralled with nets by villagers. The tuna in question are undoubtedly longtail tuna (*Thunnus tonggol*), which do indeed hug the coast, and enter the bay each year.

When European mariners began sailing into the unknown, they sometimes encountered some of the large open-ocean fishes in rather unexpected ways. No doubt, some of the earliest depictions and legends of sea monsters are based on actual observations and encounters. From early times, the billfishes were held in considerable awe, if not actual fear, since they were alleged to attack, ram and even sink ships. This extraordinary claim was thought by many to be mere superstition, but indeed, many reports of 'attacks' on ships turned out to be based on fact. Bills of both swordfish and istiophorid billfishes were often found to be embedded in the hulls of wooden sailing ships when hauled onto dry land for repairs, and sections of ships' planks, including the firmly embedded bills, reside in several museums around the world.

In 1940, EW Gudger of the American Museum of Natural History published a remarkably detailed account of this phenomenon with the marvellous title *The Alleged Pugnacity of the Swordfish and the Spearfishes as shown by their Attacks on Vessels*. He records one of the earliest accounts of swordfish ramming ships by the Roman author Aelian, written in about 120 AD, as follows:

> When the weapon made not of iron,
> but of material provided by nature,
> has reached full size, the fish ventures
> to attack a vessel in the shape of its
> own beak. Certain ones boast that
> they have seen a Bithynian ship drawn

up on shore so that her bottom [or keel] could be mended ... And then they perceived the head of a *Gladius* [swordfish] fixed so fast to it by the sword that could not be drawn away by any of the men. The body was finally torn away at the nape of the head, the beak indeed being fast [in the keel] even from the point [base] of its origin.

The earliest, and remarkably vivid account of an apparent 'attack' on a ship is given by Dutch explorer William Schouten in 1625, in his account of a voyage around the world. The translation is in old English, which adds to its flavor. Schouten's ship was off the east coast of Africa when, after some commotion, the master

> ... looked out over the side of the ship hee saw the sea all red, as if great store of bloud had beene powred into it, whereat hee wondred, knowing not what it meant, but afterward hee found, that a great Fish or a Sea monster having a horne had therewith stricken against the ship with most great strength. For when we were in Porto Desire where we set the ship on the strand to make it cleane, about seven foot under water, before in the ship, wee found a horne sticking in the ship, much like for thicknesse and fashion to a common elephant's tooth, not hollow, but full, very strong

hard bone, which had entered into three plankes of the ship, that is two thicke plankes of greene and one of oken wood, an so into a rib, where it turned upward, to our great good fortune, for if it had entred between the ribbes, it would happily have made a greater hole and have brought both ship and men in danger to be lost.

In this case, because there is quite a good description of the bill, which was like an elephant's tusk, and therefore, round in cross-section, we can be certain that the fish was a marlin. Furthermore, because of the locality (western Indian Ocean) and the force of the impact, it could only have been either a large blue marlin (*Makaira nigricans*) or black marlin (*Istiompax indica*). Another 17th-century account of an 'attack' on a ship is that of John Josselyn, an English seaman voyaging for the New World:

> ... in the afternoon [of 20 June 1638] we saw a great fish called a vehuella or Sword fish, having a long, strong and sharp fin like a sword-blade on the top of its head, with which he pierced our Ship, and broke it off with striving to get loose, one of our Sailers dived and brought it aboard.

Apart from such chance encounters with large billfish, knowledge of their existence, and that of the other giant fishes of the sea, was almost nonexistent. Most early descriptions were based on specimens which had been stranded by storms, and, these specimens being too large to transport and preserve in museums, many of the early illustrations of billfish, tuna and large sharks are highly fanciful. It was not until the late 18th century that European scientists began to describe and depict at least some of these fishes with any accuracy.

Tuna have been fished on a small scale in nearshore waters of many countries for several centuries. Methods used have included handlines baited with live fish or with lures, or setting gillnets and wall nets. However, large-scale commer-

The earliest known illustration of a marlin bill which had penetrated layers of copper-sheathed ship's timbers (1835)

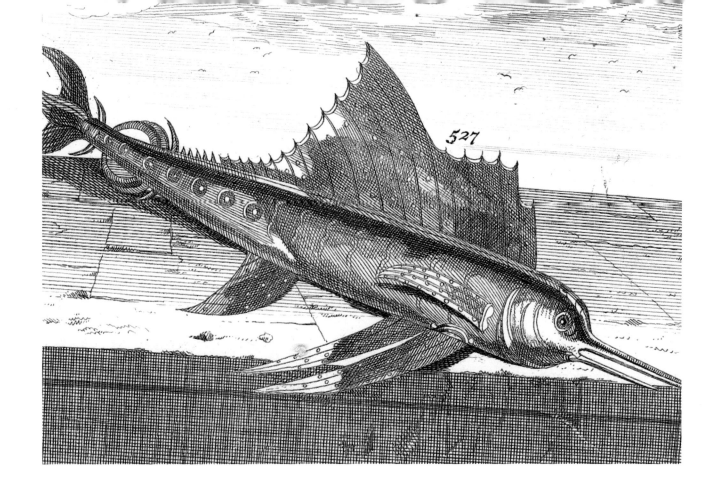

Above A fanciful Dutch illustration of a sailfish, published in 1724, suggests that the artist was not working from an actual specimen. Billfish were not depicted accurately until the mid-1800s.

Left While not usually targeted, billfish contribute additional bycatch to many coastal artisanal gillnet fisheries. These juvenile black marlin are pictured in a fish market in central Vietnam.
Julian Pepperell

29

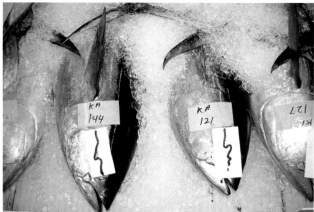

Above A large purse seiner at anchor in northern Papua New Guinea. These sophisticated vessels are capable of taking many tonnes of small tunas each time they set their nets, often covering the area of a football ground.
Julian Pepperell

Right Prime yellowfin tuna on sale at a fish house in Hawaii. Carefully killed and bled, these fish are aimed at the top-end sashimi market.
Julian Pepperell

Below right The end of the supply chain. A large tuna supplies many meals such as this array of sushi.
Dreamstime.com

cial fishing for tunas on the high seas began in earnest after the Second World War, when highly mechanized Japanese longlining fleets expanded into virgin waters of all three major oceans. These fleets began taking large catches from seemingly limitless resources, and as methods were refined, new species of fish were added to the list of catches.

Longlining, as the name suggests, involves the setting of a long main line of rope (up to 50 km in length) with up to several thousand branch lines attached, each bearing a baited hook. The automated process of setting and retrieving the gear may be repeated with rotating crews around the clock for months on end, while the catch is snap-frozen within hours of capture. While longlining targets mainly larger tuna and billfish, the smaller tunas are targeted by two other main methods: pole-and-line, and purse seine. Both of these methods are called surface fisheries, since they must locate schools of tuna at the surface before fishing commences. Pole-and-line boats throw small, live baitfish into the schools to generate a 'feeding frenzy'. A simple barbless jig on the end of a line attached to a bamboo pole is then dangled among the thrashing fish, and hauled over the shoulder whenever a fish takes the lure. If the fish are large, then two, or even three

poles joined to a single jig are used. Purse seine vessels also locate surface schools, and deploy a huge net around the fish at rapid speed. The weighted net encircles a large area, and is then 'pursed' by drawing a rope through rings at the bottom of the net to cut off escape. Over the past several decades, purse seine vessels have increased in number, targeting mainly the prolific skipjack tuna, but also, juveniles of species such as yellowfin and bigeye tunas, and in the past, southern bluefin tuna. Many fishing nations have built and commissioned their own fleets to take their share of the bountiful harvest, with the result that tunas are now the staples of huge world fisheries.

Aristotle had once regarded large, old tuna as unfit even for pickling, but times have certainly changed. At one end of the scale, top grade, large bluefin tuna, used in the sashimi (raw fish) market in Japan, can fetch as much as US$200 per kilogram, while at the other end, hundreds of thousands of tonnes of tropical tunas such as yellowfin and skipjack are canned and provide cheap protein to the world's masses.

Billfishes, in particular the broadbill swordfish, are also sought by commercial fishers. The marlins, sailfish and spearfish are generally caught as bycatch of longline fisheries, and while not a major component of the total catch, are still caught in significant quantities. The main marlin species which is caught commercially, and sometimes targeted by longliners, is the striped marlin, *Kajikia audax*, although sailfish are increasingly caught for commercial markets in the Indian Ocean.

The major tuna fisheries are among the few examples of increasing catches in world fisheries. The estimated total worldwide commercial catches of tuna and billfish combined have increased from less than 600 000 tonnes in 1950 to over 6.2 million tonnes in 2007. The bulk of this catch consists of four tuna species: skipjack, yellowfin, albacore and bigeye tuna. Although scientists currently consider that catch levels of skipjack and yellowfin tuna are sustainable, there are some concerns about stocks of other species, particularly bigeye tuna, in all three major oceans.

Even though some of the tropical tunas may be sustainably fished at present, at least two species of tuna are considered to have been severely overfished, the Atlantic bluefin tuna in the Atlantic and the Mediterranean, and the southern bluefin tuna in the Indo-Pacific. Both species are relatively slow-growing and late-maturing, characteristics which lead to low biomass turnover rates, and therefore, susceptibility to overfishing. International management regimes are attempting to redress the situation for both species, but their stocks are still estimated to be only a small fraction of the sizes they were before they were intensively fished.

The other major group of oceanic fishes which is commercially significant are sharks. Although they are not usually specifically targeted, large numbers of open-ocean sharks are caught incidentally by tuna fisheries. Baited longlines naturally catch many sharks, as do purse seine nets set around floating logs and debris. Shark fins bring a high price on Asian markets, and the crews of longliners are generally allowed to keep shark fins as a form of bonus. Apart from being seen as a waste of a whole fish, the primary problem is that the fins are often removed from live sharks, which are then dumped back into the ocean (it has been estimated that 87 per cent of longline-caught sharks are alive at the vessel when the line is retrieved). As an example, the Hawaiian longline fishery expanded rapidly in the 1990s, and fins of sharks became an important by-product of the fishery. During the period 1991–99, an annual average of 80 000 to 100 000 sharks, mostly blue sharks, were caught in the fishery and during this period, the percentage of sharks finned climbed from 10 per cent to more than 60 per cent. Internationally, over 95 per cent of sharks caught by oceanic longliners and finned are blue sharks, the most prolific shark of the open ocean. Blue sharks grow quickly and mature early, but even so, the numbers of this species are showing signs of decline in some regions.

Other species of pelagic sharks are not targeted for their fins, but for their flesh, which is often sold as 'boneless fillets'. These species are also caught by longline, and include the thresher sharks, and the shortfin mako shark. Even the largest fish in the ocean, the whale shark, is taken for food by artisanal fisheries in the Philippines, Indonesia and the Maldives.

Many other surface-dwelling fishes are also taken by commercial fisheries. These include the various mackerel (*Scomberomorus*) species, dolphinfish, wahoo, Ray's bream, opah, butterfly mackerel, escolar, oilfish, barracuda and lancetfish. Although they do not make up a large percentage of the total catch, all are important in the ecology of the open ocean.

In some counties, sport, or recreational fishing for pelagic fishes has a longer history than commercial fisheries. It is generally accepted that the first game fishing club was the Tuna Club of Avalon, formed in 1898 on Santa Catalina Island, California, by Dr Charles Holder. The formation of the club was stimulated by Holder's capture of an 83 kg (183 lb) Pacific bluefin tuna on rod and reel, the first recorded capture of a large tuna by this method. The concept of catching large, incredibly powerful fish on flimsy tackle was difficult for some to grasp, but the idea of allowing the fish a fair chance of escape soon became a popular pursuit.

The first marlins, sailfish and broadbill swordfish to be taken on rod and reel were captured in the United States and Australia in the early 1900s. By the 1920s, large specimens were being landed with tackle specifically designed for the purpose. Game fishing became very popular and adventure books describing game fishing expeditions to exotic locations became best sellers. Authors such as Zane Gray and Kip Farrington captured the imagination with their tales of battles with huge billfish, tuna and sharks in faraway places. Ernest Hemingway's allegorical tale *The Old Man and the Sea*, first published in 1952, was based on his own extensive experiences fishing for giant blue marlin in the Caribbean. The main character of the book, the

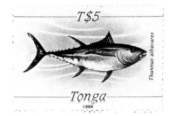

Stamps from many countries indicate the widespread importance of tuna to local economies.
Julian Pepperell

The fish that started game fishing. Charles Holder (left) with his 83 kg (183 lb) Pacific bluefin tuna caught in 1898.

poor fisherman Santiago, was based on Hemingway's regular boatman, Gregorio Fuentes.

In 1938, the International Game Fish Association (IGFA) was formed with the aim of regulating and promoting ethical angling, and keeping official records of fish captures around the globe. The principle adopted by IGFA was one originally espoused by Charles Holder:

> The underlying spirit of angling should be that it is a sport in which the skill of the angler is pitted against the instinct and strength of the fish, and that the latter is entitled to an even chance for its life.

Today, the members of thousands of game and sport fishing clubs around the world fish to this ideal.

Sport and game fishing has evolved from the pursuit of 'big game' to becoming an activity which is highly conscious of the need for conservation – so much so that the great majority of marine gamefish are now tagged and released by anglers all around the world. The origins of fish tagging may be traced to landholders in England in the 18th century, who tied colored ribbons around the tails of trout to claim ownership of fish in their streams. When fisheries science needed to plot movements and capture rates of fish, the search for a suitable means of attaching an identifying mark to fish was begun. At the end of the 19th century, scientists in the northern hemisphere were tagging plaice and sole with ivory discs fastened to each side of the fish by a silver wire passing through the body. Since then, fish tagging has included an amazing array of methods including tattooing, branding, dying whole fish, injecting radioactive isotopes, attaching metal clips, and of course, plastic loops, inserts and darts.

Tagging of small fish is relatively straightforward, but the tagging of large, potentially dangerous fishes, such as billfish, tuna and sharks, presented considerable logistic problems. The problem was solved in the early 1960s by an American scientist, Frank Mather III, of the Woods Hole Oceanographic Institute in Massachusetts. His ideas were simple, but effective, involving the jabbing of a small dart tag into the shoulder muscle of large fish without removing them from the water. The simple, plastic tags could be mass-produced and distributed to anglers free of charge, and the anglers would then do the rest; tagging and releasing their catch, and sending back the details on a pre-paid postcard. The system was an instant success, so much so that Mather's tag designs and methods of insertion of tags have changed very little over the years.

Today, international and national gamefish tagging programs are operated by governments and private organizations in many countries, including the United States, Canada, Australia, New Zealand and South Africa. Over the past three to four decades, hundreds of thousands of marlin, tuna, sharks and other marine pelagic fishes have been voluntarily tagged and released by recreational anglers. As well as instilling a strong conservation ethic in participants, the results of this massive tagging effort have been extremely valuable in helping to understand the movements and growth rates of these fishes, and hence, to aid in their conservation.

As mentioned, apart from broadbill swordfish, billfish have always been of relatively minor commercial importance compared with tuna. However, for many sport and gamefish anglers, the capture and tagging of a marlin or sailfish is the pinnacle of their fishing activity. For that reason, recreational anglers have focused attention on these species, and generated the impetus for ongoing research into their life history and biology so that their future conservation might be ensured.

Charter fishing for billfish has become a form of ecotourism which attracts export dollars to many developing island and coastal countries in all three of the major oceans. As an alternative to exploiting transient populations of tuna and baitfish for low commercial gain, many island nations are now encouraging the development of sport fishing tourism, with the added bonus of emphasizing tag and release. In many coastal towns around the world, the leaping marlin or sailfish 'logo' adorns welcome signs, local businesses, restaurants, marinas and festivals. It has become a symbol of excitement and of the healthy state of the oceanic environment.

Part two
GUIDE TO THE FISHES

5
BILLFISHES

BLUE MARLIN
Makaira nigricans

OTHER COMMON NAMES

France	Marlin bleu
Hawaii	A'u
Japan	Kurokajiki
Spain	Marlin azul
	Aguja azul

The blue marlin is a true giant of the ocean. Its confirmed maximum size of 820 kg (1805 lb) makes it the second largest teleost (bony) fish, after the oceanic sunfish, *Mola mola*. Blue marlin occur in all three of the world's major oceans. The Indo-Pacific populations have always been regarded as a single species, but blue marlin of the Atlantic and the Indo-Pacific have long been considered by some scientists to be separate species – *Makaira nigricans* in the Atlantic and *Makaira mazara* in the Indo-Pacific. The only feature described as different between the two is the pattern of the lateral line. On the surface, the lateral line of blue marlin is virtually invisible, but forms a 'chicken-wire' like pattern on the underside of the skin, the 'mesh' of which is smaller in Atlantic fish compared with those from the Indo-Pacific. However, more recent genetic studies indicate that these differences are insufficient to warrant their separation into separate species.

The blue marlin can be identified by its relatively high, pointed first dorsal fin (usually about two-thirds the maximum body depth), its folding pectoral fins and its proportionally large first anal fin. Other features are its relatively short lower jaw, and its gunmetal blue-gray color after death. A further feature to separate blue marlin from black marlin is that in the blue, the second dorsal fin is anterior to the second anal fin, whereas in the black marlin, the relative fin positions are the other way round.

The distribution of blue marlin is circumtropical, extending in summer to about 45° latitude in both hemispheres. Blue marlin are the most tropical of the istiophorid billfishes, as well as the most oceanic. They are normally found near islands or in the open ocean throughout their range, not usually being strongly associated with continental shelves (as is the black marlin).

The main recreational fishing areas for blue marlin around the world tend to be near islands. These include Hawaii in the Pacific, Mauritius in the Indian, the Caribbean islands in the Atlantic as well as many smaller islands in each ocean.

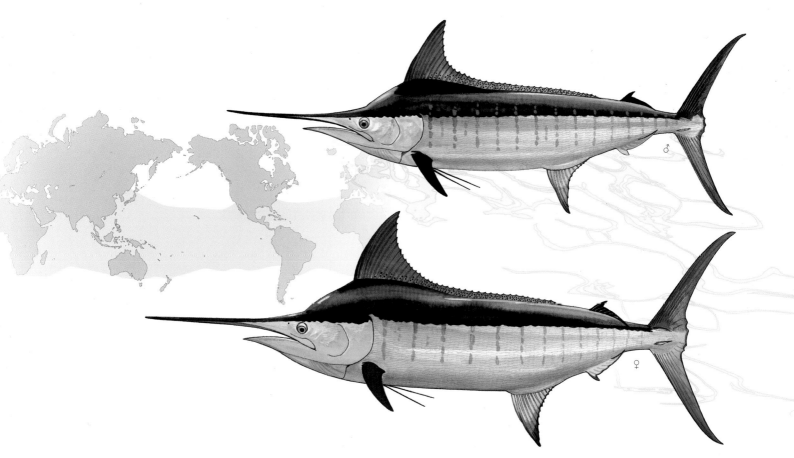

These islands are all tropical to subtropical, and appear to be locations of seasonal spawning for blue marlin.

Hawaii and Mauritius are well-known 'hot spots' for the occurrence of large, probably spawning female blue marlin, but in the Atlantic, relatively few very large – 454 kg (1000 lb) plus – specimens had been recorded until the 1990s. However, within several years of the discovery of large blue marlin around the island of Madeira, more thousand-pound blue marlin had been caught from those waters than had been taken during the long history of game fishing in the entire Atlantic Ocean. These fish were found to be not in spawning condition, and over time, their occurrence near Madeira has proven to be as enigmatic as it is unpredictable.

The majority of blue marlin have been tagged by recreational anglers in the Atlantic Ocean, mainly from the eastern United States and in the Caribbean. Over 30 years, tens of thousands of blue marlin have been tagged in the region, and hundreds recaptured. Movements of tagged fish have been extensive, with several transatlantic crossings being recorded. The most startling recapture, however, was a fish tagged off South Carolina and recaptured three years later near Mauri-

tius in the Indian Ocean – the first proven movement between two oceans for any billfish (Atlantic to Indian).

In the Pacific, tagging of blue marlin off southwestern United States and Hawaii has revealed some very extensive movements as well. The most notable of these was a blue marlin tagged off Hawaii that was recaptured near Taiwan. Off Australia, even though over 5000 blue marlin have been tagged and released, there have only been 17 recaptures to date. Again, however, one of these was highly noteworthy. This was a blue marlin tagged off the Australian southeast coast, and recaptured 18 months later by a Japanese longliner 300 nautical miles south of Sri Lanka – the second interoceanic movement of a pelagic fish (Pacific to Indian) and again, a blue marlin.

The two largest billfish ever weighed and verified, both caught off Hawaii, were blue marlin. The largest weighed 820 kg (1805 lb) and was landed by a party of novice anglers aboard a charter boat off Honolulu in 1971. The second heaviest billfish weighed 753 kg (1656 lb) and was caught off Kona, Hawaii, in 1984. There are many anecdotal reports of blue marlin weighing in excess of 2000 lb (909 kg) – even as high as 2600 lb (1180 kg) – being caught by commercial longline ves-

Above This shot of a large blue marlin epitomizes the power and beauty of this extraordinary fish.
Guy Harvey

Below A rare photo of a large blue marlin underwater, taken near the island of Madeira. Note in particular the high dorsal and anal fins.
Guy Harvey

Above A typical small blue marlin leaps in the calm waters of New Britain, off the coast of Papua New Guinea. The blue marlin is the dominant billfish species around tropical islands.
Julian Pepperell

Below Illustrations of larval blue marlin. While all other billfish develop their bills very early, blue marlin lack a bill until they attain a length of about 1 metre.
From Gehringer, JW (1956), Fisheries Bulletin (US) 57: 123–132.

sels, but it is not possible to verify any of these accounts.

There have been several studies of the growth rates of blue marlin, based on examination of their bony parts, in particular, fin spines and otoliths (ear bones). These studies have tended to assume that rings or other regular marks each represent one year of growth, but this has not been verified for blue marlin. One study which assumed annual growth rings estimated a 70 kg (154 lb) male blue marlin to be six years old; however, if blue marlin grow at a similar rate to black marlin (and they may not), then a fish of this size would only be about two to three years old.

By examining what are assumed to be daily growth rings on otoliths of very small fish, the early growth rate of blue marlin does appear to be very rapid, reaching 30 kg (66 lb) within the first year. It is theorized that all of the billfishes initially grow in length very quickly, presumably in order to be able to outswim predators as soon as possible. This is also an explanation for why very small billfish are rarely caught – it may

simply be because they don't stay small for very long.

According to early Japanese studies, blue marlin apparently have extensive spawning areas throughout the tropical and subtropical Pacific, Indian and Atlantic oceans. This picture is based almost entirely on the occurrence of larval blue marlin over very broad areas of these three oceans, and also assumes that billfish larvae have been correctly identified in these surveys.

Spawning adult blue marlin do appear to be patchily distributed, and are mainly found in seasonal aggregations around isolated islands. Around Hawaii, it has been found that nearly 80 per cent of the blue marlin landed during summer are males, mostly less than about 90 kg (200 lb) in size. Female fish caught at the same time cover a much greater size range, from as small as 22 kg (48 lb) to over 600 kg (1320 lb). Examination of the gonads of these fish has shown that spawning of blue marlin, at least at the latitude of Hawaii, is highly seasonal and predictable. Similar results have been obtained at Mauritius, indicating that

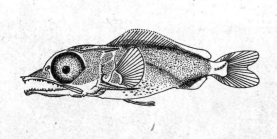

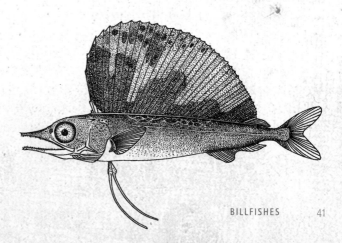

A blue marlin
about to be
released off the
Pacific coast of
Guatemala, after
being tagged
with a popup
satellite tag
Julian Pepperell

blue marlin are apparently able to locate remote islands as the spawning season approaches.

A 400 kg (880 lb) blue marlin may produce as many as 150 million eggs, each a little over 1 mm in diameter. As is the case for other billfishes, fertilization is external, and hatching of the tiny larvae occurs at the surface within two days. One recent study suggests that female blue marlin (as well as sailfish and white marlin) may spawn as many as four times in one season, but it is also speculated that spawning frequency may be much higher.

As blue marlin larvae grow, they differ from all other billfishes in that they do not develop a bill until they reach quite a large size – over 5 kg (11 lb) and 1 metre in length. This feature suggests that all of the other billfishes (broadbill excluded) form a relatively closely related group, with blue marlin representing an offshoot or outlier species.

The first attempts to track billfish after release, using a tag which emitted sonic pulses, were carried out on blue marlin off Hawaii in the early 1970s. Equipment failure and shark attack took their toll, but several successful tracks were obtained, paving the way for the more sophisticated technology of the next

two decades. In the next study, six blue marlin, all caught by rod and reel, were tracked off Kona, Hawaii, for up to 48 hours. For the first few hours after release, vertical movements were somewhat erratic; however, fish then tended to settle into relatively predictable behavior patterns. Cruising speed averaged about two knots, and the great majority of time was spent within the so-called 'mixed layer' to a depth of about 80 metres, the level of the thermocline. Thus, vertical movements of the marlin oscillated between the surface and the thermocline, with occasional brief forays into colder, deeper water. This was the first proof of the highly surface-oriented behavior of any marlin species, and has proven to be very important in understanding catch rates by commercial longlining fleets using hooks which are set above and below the thermocline.

Perhaps the most interesting finding which emerged from this, and subsequent work using more sophisticated electronic tags, is that the depth at which blue marlin swim is strongly influenced by time of day. During daylight hours, fish tend to swim deeper, often near the thermocline, whereas at sunset, behavior changes and they become much more surface-oriented during the night. What is unknown at this stage is whether feeding still takes place at night. Use of popup satellite tags over much longer periods showed that vertical behavior remains relatively predictable. There is a tendency for fish to keep moving away from their release points, but parallel to the coast, throughout the tracking periods, one fish moving over 160 km away from its release point in four days.

Like most of the other istiophorid billfishes, the blue marlin is an incidental bycatch of the tuna longline fleets of the world. In addition, significant numbers are also taken as a bycatch of purse seine vessels which set their nets around floating logs. Stock assessment studies in the Atlantic Ocean strongly point to over-exploitation of blue marlin in that region, but in the Pacific, it appears that blue marlin stocks may be stable.

BLACK MARLIN
Istiompax indica

OTHER COMMON NAMES

France	Marlin noir
Hawaii	Silver marlin
Japan	Shirokajiki
Spain	Aguja negra

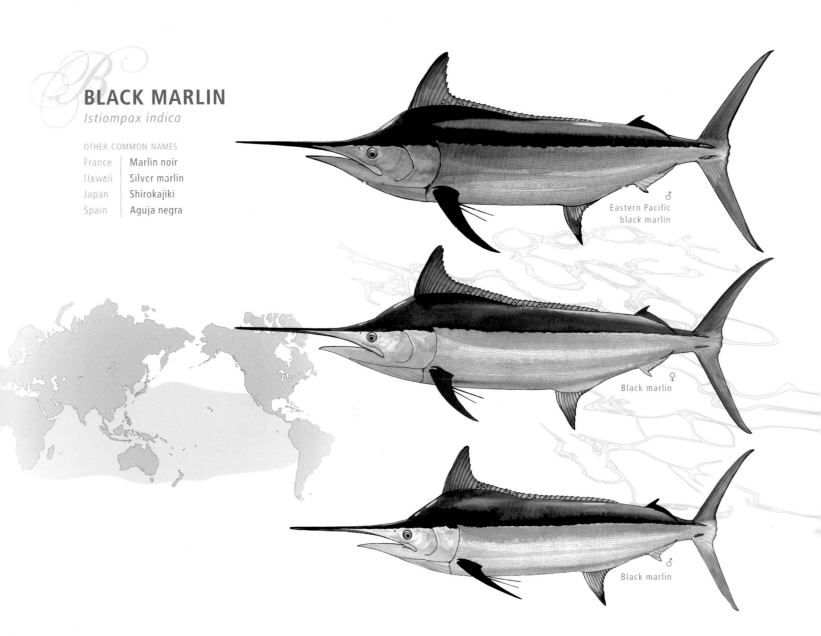

Eastern Pacific
black marlin ♂

Black marlin ♀

Black marlin ♂

The black marlin is the least common of the world's four species of marlin and, as a result, is one of the least studied and least understood of the billfishes. This is one of the largest of the teleost (bony) fishes in the world, growing to over 4 metres in length and up to 709 kg (1560 lb) in weight.

The main feature which sets the black marlin apart from all other billfishes is its rigid pectoral fins. In adults, these fins cannot be folded against the body, even with reasonable force. It should be noted, however, that very small fish have flexible pectoral fins, the calcification of the pectoral joint which causes this rigidity not occurring until a size of about 10 to 15 kg (20 to 30 lb). Another diagnostic feature is the position of the second dorsal and anal fins. The black marlin is the

only istiophorid in which the second dorsal fin is anterior to the second anal fin – a feature which holds for all sizes. Lastly, the dorsal fin of the black marlin is the lowest of all of the istiophorids, measuring no more than half the maximum body depth in adults.

Although black marlin are distributed throughout the Indo-Pacific between about latitudes 40°N and 40°S, closer examination of historic Japanese catch rates clearly shows that the density of the species is very sparse in open-ocean areas, but much more 'clumped' near large land masses and continents. In fact, the black marlin is the most land-associated of the billfishes, preferring waters on or near continental shelves during most stages of its life cycle.

Areas where black marlin aggregate include the northern part of the Great Barrier Reef, the east coast and the northwest shelf of Australia, extending to the southern islands of Indonesia, the South China Sea off Vietnam, Malaysia and Thailand, in the eastern Pacific off Peru and Central America (Panama and Ecuador) and off Kenya and Mauritius in the Indian Ocean.

Black marlin do not occur in the Atlantic Ocean; however, Japanese research longliners historically recorded stray black marlin from time to time in the Atlantic as far north as the coast of

Brazil and even in the Caribbean. The likely route for these infrequent 'invasions' would be around the Cape of Good Hope, and most world distribution maps of black marlin show dotted arrows following this route. Even so, such occurrences are considered very rare.

Most tagging of black marlin by recreational anglers has been undertaken off eastern Australia. The first black marlin was tagged off Cairns, Australia, in 1968 and since then, more than 46 000 black marlin have been tagged off eastern Australia. Of the total tagged, more than 350 recaptures have been recorded, adding

Above A black marlin shows its speed and power before being brought to the boat and tagged.
boyceimage.com

Below left A beautiful baby black marlin caught and released off southeastern Australia. Note the extremely high dorsal fin, typical of all very young billfish.
Tom Buxton

substantially to our knowledge of this species.

The long-distance movements of some tagged fish give the impression of mass long-distance dispersal of black marlin. However, this picture does not necessarily mean that many or most fish take these routes in all or most years. It does, however, clearly indicate that the species is capable of very extensive movements and that exchange of individuals throughout the species' Pacific-wide range can and does occur. A study of the genetics of black marlin in the Indo-Pacific confirmed this picture in that no differences could be found between DNA 'fingerprints' from fish taken from throughout the black marlin's range. This finding of apparent widespread mixing indicates that there is almost certainly only one stock of black marlin in the Pacific (quite possibly extending to the Indian Ocean as well). The important implication of this finding is that international cooperation in managing the species needs to be on an ocean-wide scale.

Another question raised by the tagging results is whether the movements of black marlin are random, or part of some purpose-driven migratory cycle. Tagging results clearly show that, after a period of several months, the average distance moved by tagged fish increases rapidly

with time at liberty, at least for the first nine months or so after release. This rapid dispersal takes fish away from the tagging grounds off eastern Australia throughout the western Pacific and beyond at an average rate of approximately 20 km per day. (Minimum movement rates of over 60 km per day have been recorded for some recaptured fish, but these have been short-term recaptures and are the exception to the rule.) In addition, the recapture data also show that there is a very marked 'cluster' of recaptures near the point(s) of release after about one year (330 to 400 days) followed by another period of apparent rapid dispersal in the ensuing months. Clustering of recaptures near release points is also apparent after two years, and also after three, four and five years (with decreasing numbers of recaptures as time increases). This fascinating finding suggests either that some fish never leave the areas in which they were tagged, or that annual homing occurs, at least for a proportion of the population. Careful examination of Japanese catch data for the Great Barrier Reef over long periods showed that, by early summer each year, catch rates of black marlin suddenly declined dramatically, indicating a sudden *en masse* departure of fish from that area over a very short time. Long-term charter boat captains in

This black marlin, estimated at about 10 kg (22 lb) in weight, would be less than one year old.
Peter Bristow

the area also attest to the fact that black marlin virtually disappear completely at this time, leading to the conclusion from tag returns that at least some fish must be returning to the reef on an annual basis. An important question is, what proportion of the population completes this annual cycle? Perhaps the next generation of high-tech satellite tags will provide the answer.

By analyzing the size 'pulses' of small black marlin which appear along the east coast of Australia, it has been estimated that they reach a size of about 25 kg (55 lb) at one year of age, and that

a 100 kg (220 lb) fish would be three to four years old. A small sample of very small black marlin have been aged by counting presumed daily rings on their otoliths. Previous work on tuna indicates that these rings are laid down every day during the early life of fish, and assuming that this is also the case for marlin, two 4 kg (9 lb) black marlin were estimated to have been only about 130 days (four months) old when they were caught. A rare, even smaller specimen of a black marlin, only 45 cm in length, was also aged in this way and estimated to be about 80 days old.

After several years of age, the growth rates of black marlin become more difficult to assess, but all evidence so far points to continued rapid growth. It is quite likely (but as yet, unproven) that male black marlin grow more slowly than females and die at an earlier age, explaining why all fish over about 170 kg (375 lb) are females.

The maximum size to which black marlin grow is of the order of 700 kg (1540 lb), the all-tackle world record for the species being 708 kg (1560 lb) for a fish caught off Peru in 1953, while another weighing 691.7 kg (1525 lb) was caught in the same area in 1954. Nearly 600 fish weighing more than 454 kg (1000 lb) have been captured since then, nearly all off Cairns, Australia, but none has exceeded these two long-standing records. There are persistent anecdotes of much larger black marlin being caught by longline vessels, but none of these has ever been substantiated.

Examining the gonads of black marlin caught by both recreational anglers and Japanese longliners, together with some records of occurrence of black marlin larvae, has revealed that spawning takes place in the Coral Sea in late spring/early summer each year. Egg counts from ovaries taken from adult females weighing between 400 kg (880 lb) and 600 kg (1320 lb) ranged from 65 million to 250 million eggs. As is the case for other istiophorids, the frequency of spawning for black marlin is not known.

The fully ripe eggs of the black marlin, about 1.3 mm in diameter, are fer-

tilized externally, after which they float at the surface for several days before hatching into tiny larvae. The larvae are themselves miniature predators of the planktonic world – all eyes and mouth, with one purpose – to eat and, therefore, to grow. The mortality rate during these critical early stages must be prodigious, as a whole suite of slightly larger predators take their relentless toll. Although mortality rates are obviously extremely high, it is still difficult to understand why very small black marlin, less than about 10 kg (22 lb), are extremely rare in recreational and commercial catches. It is possible that, during this phase of their life cycle, very small fish remain offshore in the midwater zone, and are therefore not available to most fishing gears, although this explanation obviously needs to be tested.

The vertical behavior of black marlin has been revealed by tracking using ultrasonic tags, as well as by analyzing data from popup satellite tags. In most cases, tagged black marlin tended to swim closer to the surface during the night compared with the day. There is also a tendency for fish to dive to deeper depths after dawn, and to make more ascents to the surface after about noon. Tagged fish rarely penetrated the thermocline, and then only briefly, remaining at temperatures no more than 8°C below that of the surface waters. The deepest dives so far recorded have only been to about 180 metres. During tracking, fish tended to initially move offshore from the edge of reefs before heading parallel to the shore. The average mean swimming speeds over the ground for tracks lasting up to 28 hours ranged from about 1.5 to 4 knots.

Although black marlin are not now targeted by most commercial fisheries in the Pacific, the numbers taken each year as bycatch are quite large. It has been estimated that in the western and central Pacific, at least 30 000 black marlin are taken annually. Beginning in the early 1950s, Japanese longliners consistently fished off northern Queensland, Australia, and during the peak years of the 1960s, up to 14 000 fish were taken annually by that fishery. A long-standing charter fishery for black marlin off the Great Barrier Reef, Australia, has operated successfully for more than 40 years. During that time, strike rates have fluctuated considerably, but the fishery has proven itself to be sustainable over this entire period. Black marlin are also a significant component of a recreational fishery off Panama. In the 1950s, many large black marlin, including the two largest on record, were caught by a small group of anglers operating from a fishing lodge in Peru. The question of whether or not that area still attracts and holds a population of black marlin is one which many would like to answer.

STRIPED MARLIN

Kajikia audax

OTHER COMMON NAMES

France	Marlin rayé
Hawaii	A'u
Japan	Makajiki
Spain	Aguja rayado

The striped marlin is a beautifully stream-lined fish with a high dorsal fin and slender bill and lower jaw. In life, the striped marlin is one of the more spectacular denizens of the open ocean. It nearly always shows vivid vertical bars, which, when feeding, can change color and intensity with remarkable rapidity. Most of the billfishes show vertical bars at least some of the time, but the striped marlin often reverses its pattern, sometimes showing dark purple stripes on a pale blue-silver background, changing instantly to phosphorescent blue stripes on a black-purple background.

Three main features distinguish the striped marlin from other billfish. Firstly, the height of the first dorsal fin is equal to or slightly less than the greatest body depth (dorsal fins of blue and black marlin are two-thirds to half the body depth). Secondly, the ratio of the lengths from the tip of the lower jaw to the caudal fork and the rear of the eye orbit to the caudal fork lies between 0.83 and 0.86, reflecting the relatively long lower jaw of striped marlin compared with blue and black marlin. This latter feature is particularly useful in distinguishing the striped marlin from the blue marlin, in which the ratio lies between 0.86 and 0.89. Additionally, striped marlin are more slender, or laterally compressed, than blue and black marlin, such that the weight of a striped will nearly always be less than a blue or a black of the same length.

The striped marlin is distributed throughout the Indian and Pacific oceans. In the Pacific, its distribution has a curious 'horseshoe' shape, with an apparent gap in the central western Pacific. Through analysis of size data from Japanese catches, scientists originally thought that there was one Pacific-wide stock of striped marlin. More recently, however, genetic studies on fish collected from widely separated sites around the Pacific have shown that there are at least four separate stocks of Pacific striped marlin – one in the eastern Pacific, one in the south central Pacific, one in the southwestern Pacific and the fourth in the northwestern Pacific. Adding weight to this finding is that these stocks each correspond with a known discrete spawning ground of the species.

Striped marlin prefer cooler temperatures than the other istiophorid billfishes, showing a preference off southern California, for example, for surface temperatures of 21–26°C. This also seems to be the preferred range for striped marlin in the western Pacific, although they are

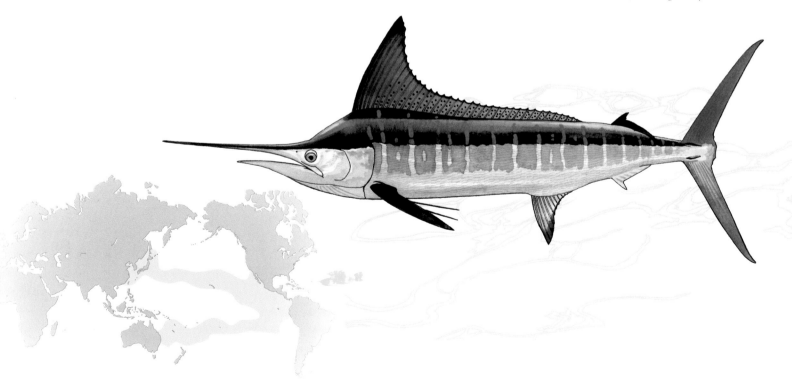

A striped marlin
showing its
vertical bars and
reflective pectoral
fins as it attends
to a packed school
of sardines. Note
the seal coming in
for a share of the
'catch'.
seapics.com/
Mark Montocchio

sometimes caught in much warmer water. In 'La Niña' years, when the western Pacific warms more than normal, striped marlin are sometimes caught as far south as the northeast coast of Tasmania, Australia. In those years, it is at least theoretically possible that striped marlin may move between the Pacific and Indian oceans via the Southern Ocean.

Many thousands of striped marlin have been tagged by recreational anglers off southern California and Baja California. Some recaptures of tagged fish have demonstrated lengthy movements as far away as Hawaii and the Marquesas, but the majority of tagged fish have been recaptured within a few hundred miles of their release points. Striped marlin have also been tagged in good numbers off northern New Zealand, also indicating some site fidelity, but with a number of extensive movements as far as Fiji, and in one case, the Marquesas Islands. Recaptures of striped marlin tagged off eastern Australia have mainly indicated relatively short, coastal movements, with some longer distances recorded, but all within the boundaries of the Coral Sea and New Zealand.

As a result of these tagging studies, it is now becoming clear that striped marlin do not undertake as extensive movements as black and probably blue marlin. The tagging results also support the conclusions of the genetic studies, substantiating that there are separate stocks of striped marlin on either side of the Pacific Ocean.

Striped marlin probably grow relatively quickly, although few measurements of their growth rate have been attempted or validated. It has been estimated that striped marlin reach about 90 kg (200 lb) by their third year of life, and that their maximum life span may be about 16 years. Because knowledge of a species' growth rate is essential for understanding its ability to respond to fishing pressure, it is very important that definitive studies on striped marlin growth rates are undertaken in the near future.

Striped marlin do not grow nearly as large as blue or black marlin, both of which may attain weights in excess of 700 kg (1500 lb). The largest striped marlin caught on rod and reel weighed 243.6 kg (536 lb) and was landed off New Zealand in 1995. This specimen was nearly 20 kg (44 lb) heavier than the official IGFA record, and because of its size, its identification was in doubt until it was able to be positively identified.

The major spawning grounds of striped marlin in the southwestern Pacific appear to be over an extensive undersea mountain chain or ridge between New Zealand and New Caledonia. Spawning is also known to occur in the eastern Pacific, the northern Pacific and the south central Pacific, as noted, adding weight to the likelihood of four distinct stocks within the Pacific basin. Larvae have also been collected in the Banda and Timor seas of the Indian Ocean, so it is likely that at least one genetic stock also occurs in the Indian Ocean. Based on the occurrence of ripe ovaries in female fish, spawning over the southwestern Pacific undersea ridge occurs throughout summer.

Adult females may contain as many as 29 million eggs at any one time during a spawning season. Pairs of fish have been

Striped marlin spend a high proportion of their time at or near the surface. Here, a small fish investigates a trolled artificial lure.
seapics.com/ Michele Hall

observed during spawning times, and recreational anglers have reported that when a striped marlin is hooked, its partner may stay close until the fish is caught or released. It has not been proven, however, that this behavior is due to pair bonding or feeding interest.

Striped marlin and the other billfishes are what are termed 'opportunistic' feeders. This means that they take the opportunity to feed on what is available at the time, although obviously within certain limits of size and palatability of prey items.

Generally speaking, billfish eat fish and squid, and the ratio of these food items in the diet probably reflects their relative availability. Striped marlin, having the coolest temperature preferences of any of the istiophorid billfishes (which excludes swordfish), tend to live in waters which are rich in squid. Of course, striped marlin may well have evolved in this direction of temperature preference to take advantage of squid resources, but whatever the reason, cephalopods make up an especially large component of the diet of striped marlin in most areas. For example, a study of striped marlin stomach contents in New Zealand found that 55 per cent of all fish contained squid, the next most common food items being jacks (24 per cent) and blue mackerel (21 per cent).

One of the characteristics of the striped marlin is that its flesh is often a marked orange color. Sometimes the color can be quite intense, even resembling smoked salmon. This sort of color is highly favored by the Japanese market, with the result that striped marlin are particularly valuable in the specialty sashimi trade. The orange color of the flesh derives from carotene, which crustaceans contain in abundance (and is why prawns, lobsters and crabs turn orange when cooked). Squid primarily feed on small crustaceans, and since striped marlin eat a lot of squid, they tend to concentrate carotene in their flesh. This also explains why not all striped marlin have orange flesh, and why, therefore, this characteristic cannot be used as an absolute identifier of striped marlin.

Early studies of the behavior of striped marlin employed ultrasonic transmitting tags off the coasts of southern California and Hawaii. In both cases, all fish spent a very high proportion of their time close to the surface. Few tracked fish ever ventured into cooler waters below the thermocline, which, in both cases was only 80 to 100 metres below the surface. During the tracking of striped marlin off California, apparently 'normal' behavior of tracked fish was observed, including feeding, tailing and long periods of quiet behavior which were interpreted as sleeping.

More recently, electronic satellite tags have been used to study striped marlin off Mexico, Ecuador, Australia and New Zealand. Popup satellite tags were deployed on nearly 250 striped marlin spread among all these locations. The average duration that tags stayed on fish was between two and three months, and in that time there was strong evidence for broad regional fidelity. As an example, one fish tagged off New Zealand moved north, made a circuit of New Caledonia and returned to New Zealand waters eight months later. In another novel experiment, scientists attached 'spot' tags to the upper tail lobe of striped marlin released off northern New Zealand. These showed that some fish stayed in the general region, but that others moved away from release sites in seemingly purposeful directions. A number of these were noted to track directly over an undersea ridge which extends north of New Zealand towards Fiji. This suggests that the surface-oriented marlin could detect the deep underwater features, presumably via an upwelling effect.

The striped marlin is not the most common of the istiophorid billfishes in the Pacific Ocean, but it is the most valuable. As a consequence, it has become a target species in some longline fisheries. Global catches of striped marlin peaked at around 24 000 tonnes in the 1960s, but in the 2000s have only averaged about 8000 tonnes. Nominal catch rates (fish per hook) have also declined and there is some concern regarding status of the stock of striped marlin in the central and western Pacific.

Recreationally, striped marlin support active charter fisheries in many regions, including Mexico, New Zealand, southeastern Australia and Kenya.

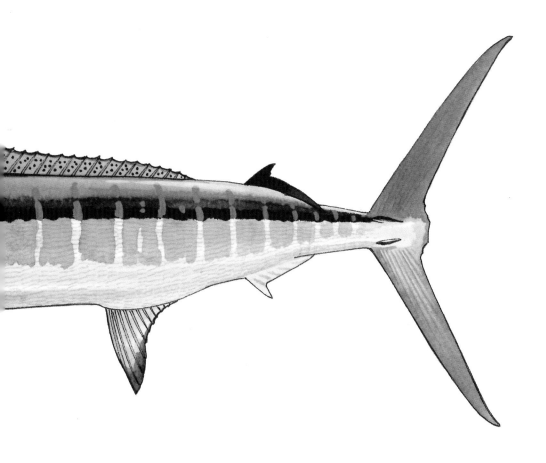

WHITE MARLIN
Kajikia albida

OTHER COMMON NAMES

France	Makaire blanc
Japan	Nishimakajiki
Spain	Aguja blanca

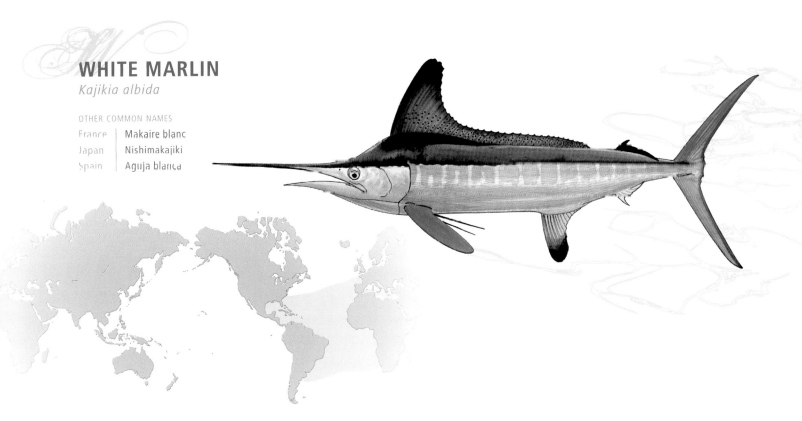

The white marlin is the Atlantic Ocean congener of the Indo-Pacific striped marlin (*Kajikia audax*), both being remarkably similar in morphology and habit. In fact, genetic differences between the two species are very slight, indicating that they split from the same ancestral stock perhaps as recently as a million years ago.

The white marlin occurs only in the Atlantic Ocean and the Mediterranean Sea. It is distinguished from the only other Atlantic marlin, the blue marlin, by its very high first dorsal fin (as deep or nearly as deep as the body), its relatively longer lower jaw, and finer denticles on its bill. It is also widely noted that white marlin have quite rounded tips to their first dorsal and first anal fins. While this is probably the case generally, specimens with quite pointed dorsal and anal fins are common enough to cast doubt on this feature being listed as diagnostic. White marlin do not grow to sizes approaching those of blue marlin – in fact, rarely attaining 80 kg (176 lb). The other species with which the white marlin could be confused, especially as juveniles, are the longbill and roundscale spearfishes, *Tetrapturus pfluegeri* and *Tetrapturus georgii*. The longbill spearfish is more slender and, importantly, the anus of

both spearfish species is positioned well forward of the first anal fin. In the white marlin, the anus is close to the anal fin origin.

As noted, white marlin are confined to the Atlantic Ocean and Mediterranean Sea, between latitudes of about 45°N and 45°S. In warmer years, its range extends further, with some records of specimens caught off France, and one record of a single fish found near the coast of England. In the southern hemisphere, white marlin are occasionally taken off the Cape of Good Hope as well as off northern Argentina.

The white marlin was the first billfish to be scientifically tagged under a cooperative tagging program. Frank Mather III of the Woods Hole Oceanographic Institute coordinated the first program of this kind in the early 1960s, and showed that istiophorid billfish were good candidates for tagging by sportfish anglers. Between then and 2005, nearly 50 000 white marlin have been tagged, nearly all in the Gulf of Mexico and off the eastern United States. Over 950 recaptures have been recorded for the species, and results show that, while some individual movements may be extensive (several Atlantic crossings have been recorded), overall

movement has been almost exclusively confined to the northwestern Atlantic, with very few fish venturing east of longitude 65° and, significantly, only two fish crossing the equator. The few transatlantic movements recorded for white marlin contrast with blue marlin and bluefin tuna, both of which have recorded multiple crossings of this ocean. The longest distance moved by a white marlin is less than half that moved by blue marlin or black marlin. Times at liberty for recaptured fish have usually been less than three years, although some recaptures after ten years, and one after 15 years, have been recorded. Annual clumping of recaptures near release points strongly suggests that this species homes on a cyclic, annual basis, or else is very restricted in its general movements.

The growth rate of white marlin is not well understood, but is probably rapid, with fish reaching about 30 kg (66 lb) by two years of age. The maximum size of white marlin is 82.5 kg (181 lb) – a fish caught off Brazil in 1979. Most of the fish which have weighed more than 70 kg (154 lb) have been taken off Brazil.

As noted above, a handful of white marlin have been recaptured ten to 15 years after tagging, indicating that some individuals live at least that long. However, it is generally thought that this is a relatively short-lived species, with the bulk of the population not living beyond ten years.

White marlin are late spring/summer spawners in both the northern and southern hemispheres. Off Brazil, as evidenced by the appearance of large adults, larvae and postlarvae, they spawn from November to April. In the Caribbean, on the basis of gonad development, the spawning season appears to be April to June. This is in keeping with the observation that summer concentrations of white marlin found in the Gulf of Mexico and off the southeastern United States are feeding, spent (spawned out) fish, implying that in this area, spawning probably occurs in spring.

There have been several dietary studies of white marlin, all of which have shown them to be rather opportunistic predators, eating a wide range of fishes and invertebrates, but with a preference for squid in particular. This preference is similar to the striped marlin, and may go part of the way to explaining why the flesh of both species is usually orange to salmon pink in color. It is likely that this coloration derives from carotene which squid have accumulated from eating small crustaceans in the water column, transferring this pigment to the white marlin where it colors the flesh in much the same way that salmon and trout flesh becomes colored through eating aquatic crustaceans directly. As well as squid, other items found in white marlin stomachs include dolphinfish, herring, various trevallies (jacks), lancetfish, cutlassfish, barracuda, triggerfish, crabs and even small hammerhead sharks.

Even though it is a small billfish, the white marlin is a very important recreational species in a number of countries in the western Atlantic, in particular, the United States, Venezuela and many Caribbean islands.

White marlin are caught incidentally by tuna longliners, the total Atlantic catch peaking at just under 5000 tonnes back in 1965. Since 1970, catches have averaged about 1500 tonnes while total fishing effort has increased considerably. It is now estimated that the current stock size of white marlin in the Atlantic is at a level of between 5 per cent and 15 per cent of the optimum carrying capacity. Stocks have been regarded as being well below the sustainable yield level for many years. Even though the species is not regarded as being in imminent danger of extinction, it is considered that unless fishing mortality is reduced soon, and substantially, the population could be driven to dangerously low levels.

A white marlin clearly showing the long, rounded anal fin characteristic of this species. Also note the rounded dorsal fin and the long pectoral fin with somewhat rounded tip.
Guy Harvey

SAILFISH
Istiophorus platypterus

OTHER COMMON NAMES

	Bayonet fish
France	Voilier
Japan	Bashukajiki
Spain	Pez vela

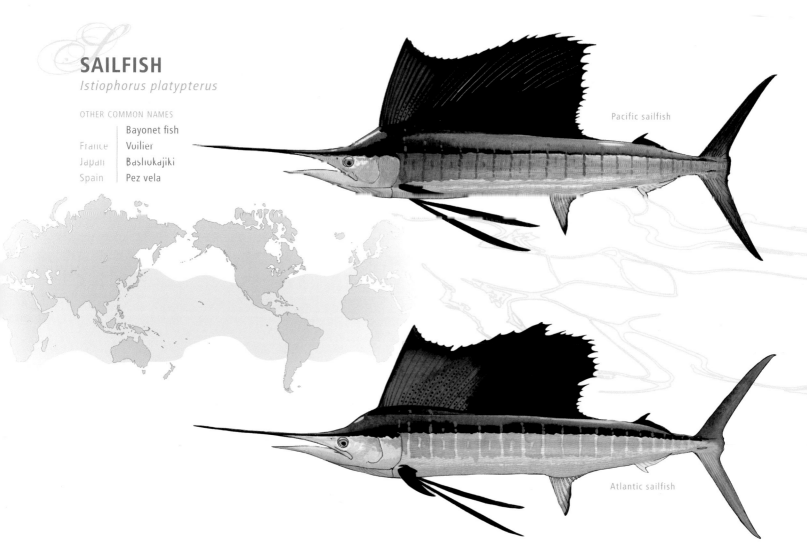

Pacific sailfish

Atlantic sailfish

The sailfish is a member of the family Istiophoridae, which also includes the marlins and spearfishes. The istiophorid billfishes are characterized by possessing a pair of lateral keels on either side of the caudal peduncle or tail wrist, round bills in cross-section and by their long-based dorsal fins. The sailfish is readily identified by its extraordinarily large first dorsal fin, by far the highest of the billfishes, leading to its name *Istiophorus platypterus*, which literally means 'flat-winged sail bearer'.

In the early part of the 20th century, sailfish of the world were divided into as many as ten species depending on where they were caught. These distinctions were often based on minor anatomical differences, especially the height and shape of the dorsal fin. The debate has now condensed into the possibility of there being just one or two species of sailfish. Some authors consider there to be two – one in the Atlantic (*Istiophorus albicans*), the

other spread throughout the Indian and Pacific oceans (*Istiophorus platypterus*). However, as indicated below, it is now accepted by most authorities that there is only one species of sailfish around the globe, *Istiophorus platypterus*. This does not mean that there may not be some regional genetic variations, but just that such differences are not large enough to be classified as different species. The two illustrations here show typical Atlantic and Pacific sailfish, bearing in mind that they are the same species.

Worldwide distribution maps of the sailfish convey the impression that the species is spread right throughout the three major oceans. While this is true to some extent, sailfish are much more concentrated near land masses and islands than in the open ocean. One reason that their apparent distribution is pan-oceanic may be the Japanese, who originally mapped pelagic fish catches by their long-

Left Sailfish have by far the largest dorsal fin of any of the billfishes. This photo also shows the very long pelvic fins under the breast, with trailing membranes.
Guy Harvey

Below Tiny sailfish such as this are sometimes attracted to lights at night. This one, dip-netted off Florida, would only be weeks old.
Carey Chen

line fleet, traditionally lumped sailfish and spearfish catches together under one category. More recently, observers on longline vessels have recorded that, in the open ocean away from land, spearfish are far more numerous than sailfish. Therefore, sailfish distribution may prove to be far more 'patchy' than shown in the standard texts.

Sailfish often aggregate where nutrient-rich coastal water meets oceanic water to form fronts. These fronts are visible from satellites, and are known to be areas of concentration of plankton which in turn attract planktivorous fish such as pilchards and anchovies. And where such baitfish congregate in numbers, so too do sailfish. Often, this nutrient-rich water flows into the sea from mangrove-lined rivers, bays and inlets, so it is seaward of such habitats where concentrations of sailfish might be expected.

Even though the billfishes as a group are sometimes called 'fish without a country', referring to their tendency to move large distances, even sometimes across entire oceans, cooperative tagging programs around the world have shown that sailfish differ from the marlins in not showing strong tendencies to travel far. In the Gulf of Mexico and the southwestern Atlantic, where many thousands of sailfish have been tagged by recreational

anglers, movements have been quite limited. As if to prove the rule, there have been a small number of long-distance travelers, the record being set by a sailfish tagged off North Carolina and recaptured near the Guineas in the southern Caribbean, a distance of 3000 km. Recaptures of sailfish tagged in Australia on both the east and west coasts also indicate that the great majority of tagged sailfish have

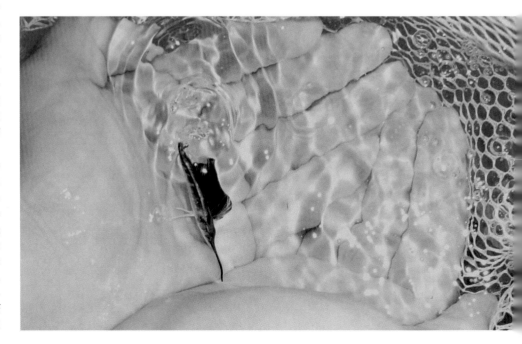

moved very little, if at all, even after times at liberty as long as four years. Again, there are some exceptions, but of nearly 250 recaptured fish, very few had moved more than 200 km. Finally, a study in the Persian Gulf, in which sailfish were tagged using both conventional and electronic tags, found that the population there is entirely confined to that relatively small body of water.

For many years, sailfish were considered to be a particularly short-lived species, perhaps only living for four or five years. Growth rates based on size frequency analysis and lack of any tag returns after four years seemed to support that view. However, in 1984 a tagged female sailfish was recaptured in the Atlantic after being at liberty for 4025 days (10 years, 10 months). This recapture completely changed these earlier notions and we now realize that, even though this particular fish may have been a slow grower, some individual sailfish probably do live into their teens.

The most recent study of the growth rate of sailfish was conducted in Taiwan, where scientists and students at the National Taiwan University have access to thousands of specimens of billfish at the central fish market in Shinkang. The dorsal fin spines from over 1100 sailfish were sectioned and the visible growth rings counted to estimate age. This study indicated that sailfish, at least from near Taiwan, may have slower growth rates than other studies had suggested. Fish of about 20 kg (44 lb) were estimated on average to be four years old, while a fish of 35 kg (77 lb) might be as much as eight years old, although there is a wide range of sizes for any given estimated age. Interestingly, this study showed that female fish reach a larger size than males, and confirmed the suspicion for all of the istiophorid billfishes that females also grow faster than males.

Sailfish grow to different maximum sizes in different oceans, and appear to reach different maximum sizes even on different sides of the same ocean. It has long been known that sailfish tend to be much larger in the eastern Pacific (off Ecuador and Mexico) than in the west-

ern Atlantic (off Florida and Brazil). In the 1980s, a study of Japanese longline data showed that sailfish in the eastern Atlantic off Africa were also relatively large. Checking the record charts of IGFA reveals that nearly all of the Atlantic records for sailfish were indeed caught off the west coast of Africa, mostly in the 1990s. The record for the Atlantic stands at 64 kg (140.8 lb), caught off

Angola in 1994. In contrast, all line class records for Pacific sailfish are greater than 60 kg (132 lb), the current all-tackle record being 100.2 kg (221 lb), caught off Ecuador in 1947. Most of the other line class records were caught in the eastern Pacific, although large sailfish over 90 kg (198 lb) have been caught in both the Philippines and Tonga. There are no current IGFA line class records for sailfish from the Indian Ocean, but the largest sailfish weighed from Western Australia was 77.9 kg (171 lb). The variations in maximum and average sizes of sailfish in

Sailfish cooperatively feeding on a 'ball' of baitfish. The huge dorsal fins are raised like capes to keep the ball tight, and individual fish then take it in turns to move in for a feed.
Guy Harvey

Moving off with the prize. This sailfish holds its prey in its mouth prior to crushing and swallowing it. The jaws of billfishes have no teeth but are very powerful.
boyceimage.com

different areas may be due to differences in environmental conditions or to differences in genetics at a population level.

The ovaries of a mature sailfish of about 35 kg (77 lb) may contain as many as five million eggs. Spawning is thought to take place in relatively shallow water near land masses, or near islands. Spawning seasons vary, but may occur throughout the year with a peak period during summer months. Fertilization is external and the eggs, about 1 mm in diameter, hatch into larvae within several days. The larvae bear little resemblance to adult sailfish, but are nevertheless ferocious little predators of the warm surface layers of the ocean. Quantities of sailfish larvae up to about 10 mm long have been collected during larval surveys, but they are rarely seen above this size. This is because the larger larvae and postlarvae are mobile enough to avoid towed plankton nets. The most effective way of catching these miniature sailfish is by dip net, usually at night under a bright light. Very small billfish are extremely valuable scientifically, so if one is caught or found in the stomach of another fish, every effort should be made to freeze or preserve it.

Size at first maturity seems to vary considerably with location, and size is not necessarily correlated with maturity. In a Taiwanese study, the smallest mature female fish measured 162 cm (lower jaw to caudal fork), which equates to about 20 kg (44 lb) in weight. My observations have shown that female sailfish off northeast Australia are usually mature at weights of about 30 to 35 kg (66 to 77 lb), while those off northwestern Australia are mature by about 23 kg (50 lb). Again, these differences may be environmental, genetic or both.

Sailfish cooperate with each other when feeding on small, schooling baitfish such as anchovies, sardines or, surprisingly, pelagic pufferfish. The hunt appears to begin with one sailfish making a series of leaps and re-entry splashes, some-times as many as 20 consecutive jumps at a time, in a semicircular route. The purpose of this behavior, known as 'free jumping', is almost certainly to round up a school of baitfish in an area into a panicked, tight ball. Other sailfish, perhaps 10 or 20 at a time, will then move in, and may be seen swimming leisurely around and below the ball of baitfish. If the ball begins to disperse, one or two of the sailfish will immediately raise their sails like great, dark capes and flash their reflective blue and purple colors, immediately causing the baitfish to again pack tightly together. Then, one by one, the predators will casually move into the ball, striking with their bills, or simply plucking a single fish from the ball with a deft turn of the head.

Although sailfish are not a targeted commercial species by high seas longline fisheries, they are caught in increasing quantities by coastal gillnet fisheries in many developing countries, including Kenya, the Maldives, Malaysia and Vietnam. That said, there is no doubt that sailfish are a significant bycatch of the world's tuna fisheries. This also applies to purse seine fisheries which set their nets around logs, incidentally catching any attendant sailfish.

In the Indian Ocean, where sailfish are perhaps targeted more than in other major oceans, the minimum average annual catch by all countries was estimated for the period 2002 to 2006 to be around 24 000 tonnes, which, if fish averaged 20 kg (44 lb), would equate to 1.2 million fish.

By far the most significant fisheries for sailfish, however, especially in terms of economic value, are sport and charter. Sailfish tournaments generate considerable income to local economies in a number of countries, and sailfish charter fisheries in developing or resource-poor countries, such as the Maldives, Ecuador, Guatemala, Mexico, Costa Rica, Thailand and Fiji, are proving to be viable alternatives to commercial exploitation.

SPEARFISHES

Tetrapturus angustirostris,
T. pfluegeri, T. belone, T. georgii

OTHER COMMON NAMES

Shortbill spearfish *T. angustirostris*

| Japan | Fuuraikajiki |
| South America | Pez aguja corta |

Longbill spearfish *T. pfluegeri*

| Japan | Kuchinaga |

Mediterranean spearfish *T. belone*

France	Poisson pique
Italy	Aguglia imperiale
Japan	Nishifuuraikajiki
Malta	Pastardella

Roundscale spearfish *T. georgii*

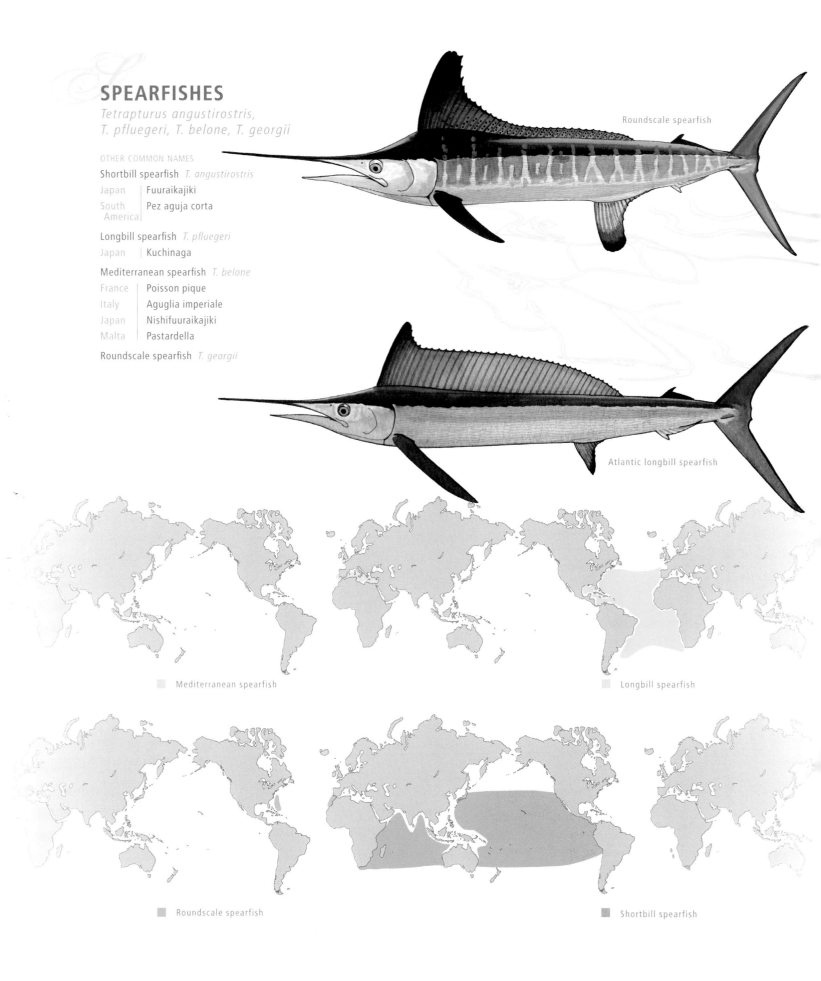

Roundscale spearfish

Atlantic longbill spearfish

Mediterranean spearfish

Longbill spearfish

Roundscale spearfish

Shortbill spearfish

The name 'spearfish' was originally bestowed on all members of the marlin/sailfish family, Istiophoridae, since the bills of all these fishes are round in cross-section (like a spear). This was in contrast to the aptly named 'swordfish' (*Xiphias gladius*), which has a flattened, sword-like bill. Today, 'spearfish' is reserved for the four *Tetrapturus* species outlined here.

The primary distinguishing feature which separates spearfish from the marlins and sailfish is the position of the anus, or 'vent'. In all spearfishes, this opening is located well forward of the base of the first anal fin (at least as far forward of the anal fin as the longest spine of that fin), whereas in the other istiophorids it is located just forward of the origin of the anal fin.

After a confusing taxonomic history, it is now accepted that there are four distinct species of spearfish. One of these, the shortbill spearfish (*Tetrapturus angustirostris*) presents no identification problems since it is the only spearfish which occurs in the Pacific and Indian ocean basins. In the Atlantic and Mediterranean, however, three species of spearfish occur, causing some problems with identification. These are the longbill spearfish (*Tetrapturus pfluegeri*), the Mediterranean spearfish (*Tetrapturus belone*), and the now confirmed roundscale spearfish (*Tetrapturus georgii*), originally recorded only in the Mediterranean and off Portugal but more recently confirmed in the western Atlantic. All of the spearfishes are relatively long, slender fish, with maximum sizes generally not exceeding 45 kg (100 lb), although as discussed further, larger specimens may occur.

In the Indo-Pacific, the only other fish with which a shortbill spearfish might be confused is a very small black marlin, since both have relatively long, slender bodies. As noted, the position of the anus will immediately identify the fish as a spearfish or a marlin; however, if this is not possible to check, the length of the pectoral fin is also a good feature. In the shortbill spearfish, this fin is very short compared with any marlin species, including the black. Lastly, even in small black marlin, the bill is long and spindle-like, but very short (only a little longer than the lower jaw) in the shortbill spearfish.

In the Atlantic Ocean, because there are three species of spearfish, their identification is not quite so simple. As its name suggests, the longbill spearfish has a 'typical' length bill for a billfish, and so can be confused with marlins, in particular the white marlin. Again, the primary distinguishing feature to separate longbill spearfish from white marlin is the forward position of the anus.

The principal feature of the roundscale spearfish which sets it apart from other billfishes is, as the name suggests, the possession of scales with soft, rounded ends (all other billfishes have stiff, pointed scales). For decades, the debate raged as to whether or not this was really a distinct billfish species or perhaps a hybrid. The issue was finally resolved in a paper published in 2005 which showed that the roundscale was indeed a separate billfish species.

The fourth spearfish, the Mediterranean spearfish, has been known for centuries to occur in the Mediterranean. It has the long lean body of a spearfish, together with the high, long dorsal fin and forward-placed anus. It is further distinguished from the other Atlantic spearfishes, and the white marlin, by its very short pectoral fin (much shorter than the strap-like pelvic fins under the breast) and its bill, which is not as long as that of the other Atlantic spearfishes, or the white marlin.

The shortbill spearfish is widely distributed throughout the Indian and Pacific oceans. It maintains a relatively narrow geographic range, between latitudes of about 35°N and 35°S. It is very much a fish of the open ocean, rarely found near continents. Although it is entirely endemic to the Indo-Pacific, in 1974 a Japanese research longliner caught three verified specimens of the shortbill in the Atlantic Ocean. These were all caught off the west African coast, presumably moving into the Atlantic via the Cape of Good Hope. Interestingly, one of these fish weighed 42 kg (92 lb), one of the largest specimens on record.

A shortbill spearfish which has just been tagged and is about to be released. Note the long, narrow body and the height of the dorsal fin through its whole length.
seapics.com/ Doug Olander

The longbill spearfish is widespread throughout the Atlantic, with an apparently broader temperature tolerance than the shortbill, being found between somewhat higher northern and southern latitudes. It was only recognized as a separate species in the 1960s, its specific name, 'pfleugeri', being in honor of the long-standing Florida-based taxidermist Al Pfleuger, who brought the fish to the attention of biologists.

As its name suggests, the Mediterranean spearfish is mainly confined to the sea after which it is named, although it is noteworthy that the IGFA record for this species was caught near the island of Madeira, about 1500 km from the entrance to the Mediterranean.

The story surrounding the 'discovery' of the roundscale spearfish is intriguing. In the late 1950s, Richard Robins, an American scientist who had heard of unusual billfish being caught off Portugal, arranged to purchase specimens of billfish caught in the vicinity of Portugal, Spain and Sicily. In 1961, he made the trip to examine 95 fish which had been collected and frozen by commercial fishermen. Of 36 Sicilian fish, 35 were Mediterranean spearfish, while 56 of the remaining 59 specimens were identified as white marlin. However, the other four

proved to be the rare species that Robins had been looking for, the so-called roundscale spearfish. The much more recent discovery, via genetics, of the roundscale spearfish in the northeastern Atlantic now indicates that its distribution is much broader than Robins thought, covering both sides of the Atlantic, and penetrating into the Mediterranean Sea.

Of several hundred shortbill and longbill spearfish tagged, there have been no recaptures of either species. While this is somewhat frustrating, we can probably speculate on general movements. The shortbill is highly oceanic and widespread, so it is likely that it wanders fairly freely over its range. Similarly, because of their widespread transatlantic distributions, it could be speculated that the longbill and roundscale spearfish are also highly mobile, in contrast to the Mediterranean spearfish, which apparently has a very restricted distribution with only limited 'leakage' outside the Mediterranean basin.

There are no known studies on the growth rate of the shortbill spearfish. In the absence of such data, it could be reasonably surmised that this is a fast-growing species, since other istiophorids are among the fastest growing of all fishes. It might also be guessed that the life span of shortbill spearfish would be relatively short, since their maximum size is the smallest of the billfishes. These suppositions are perhaps supported by the one study which did attempt to estimate ages for a spearfish species – the Atlantic longbill. Based solely on length measurements of landed fish, that study suggested that longbill would be no more than four years old at maximum size, although more direct methods of aging would be essential to be certain about growth rates and longevity.

The maximum size reached by the spearfishes is not always easy to determine. Probably the best source of data in this regard is the IGFA's record list, although in some cases, there may be issues with respect to correct identification. This is probably why, for line class records, IGFA lumps all of the Atlantic spearfishes under the mixed heading

Tetrapturus spp., presumably to avoid the problems associated with identification among the longbill, roundscale and Mediterranean species. IGFA does, however, distinguish between the longbill and the Mediterranean for all-tackle purposes. The current IGFA all-tackle records for spearfish are: shortbill spearfish 36.8 kg (81 lb), caught off northern New Zealand in 2003; Atlantic longbill spearfish 58 kg (128 lb), caught off the Canary Islands in 1999; and Mediterranean spearfish 41.2 kg (90.6 lb), caught off Madeira in 1980. Regarding the roundscale spearfish, the 1963 study of four specimens noted that the largest weighed 23.5 kg (52 lb). Unfortunately, the more recent study of 16 western Atlantic specimens did not publish sizes.

An authoritative Japanese reference lists particularly large maximum sizes for some of the spearfishes. This states that shortbill spearfish grow to 52 kg (114 lb) – note the current world angling record of 36.8 kg (81 lb) – and that the Mediterranean spearfish reaches 70 kg (154 lb) – current record, 41.2 kg (90.6 lb). In the case of the shortbill spearfish, the original reference mentions a specimen weigh-

ing 51.8 kg (114 lb) at the Honolulu fish markets in 1957, so that would seem to be a valid weight. On the other hand, the figure of 70 kg for the Mediterranean spearfish cannot be verified.

The gonads of spearfish are similar to those of the other billfishes, except that they show marked left–right disparity in size, the left ovary and testis always being much larger than the right. This has led to the description of the shape of the paired ovaries and testes in spearfish as markedly 'y-shaped', another characteristic of the group.

Presumably, spearfish produce many millions of eggs, although again, because of their relatively small size, far less than the tens or hundreds of millions produced by the blue and black marlin. Only one study has reported actual egg counts from shortbill spearfish, and only for two fish, both about 150 cm long. These showed considerable variation, one fish having 2.1 million and the other 6.2 million eggs.

Very small juveniles of shortbill spearfish have been described, and it is interesting that, at sizes up to about 40 cm in length, the shortbill spearfish

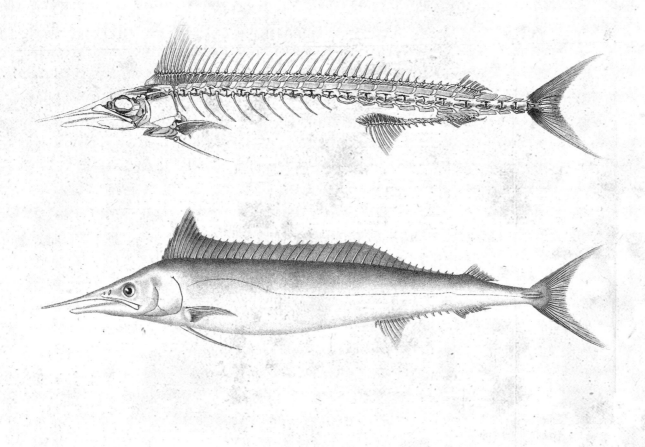

actually has a long bill. Beyond this size, the bill virtually stops growing, resulting in adults possessing the shortest bill of any billfish.

There have only been a few observations on the diets of spearfish, indicating that they eat much the same food items as the marlins (fish, squid and some crustaceans) but that their stomachs do not contain deep-water fish which are sometimes found in marlin diets. This implies that spearfish are strongly confined to the upper ocean layers, even more so than marlins.

Commercially, spearfish are caught almost exclusively as a bycatch of longlining, the great majority by the distant-water fleets of Japan, Korea and Taiwan.

Unfortunately, it is impossible to determine from historic records how many spearfish were routinely caught by these fleets because spearfish and sailfish were commonly lumped together under one category. More recently, separate records have started to be kept for all billfish, and these indicate that, at least in the open ocean, spearfish dominate the lumped sailfish/spearfish category.

As food fish, spearfish are regarded very highly. Shortbill spearfish is certainly a regular species seen in Hawaiian fish markets. None of the spearfish species appears to concentrate in predictable regions, so it is unlikely that the species could be specifically targeted for a commercial market.

BROADBILL SWORDFISH
Xiphias gladius

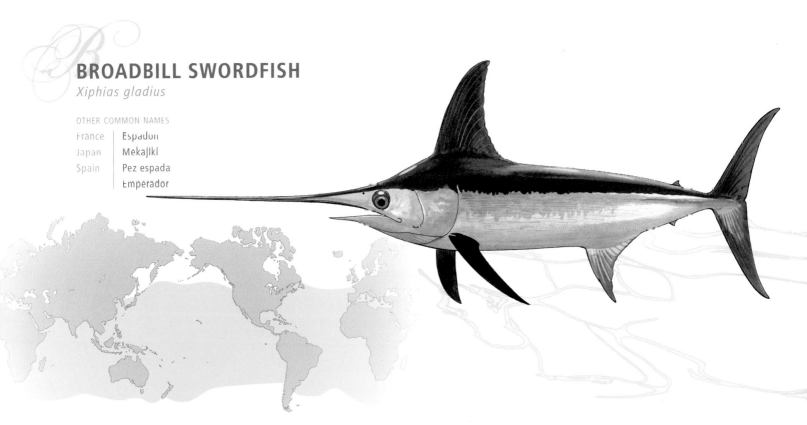

Because the broadbill swordfish was relatively common in the Mediterranean Sea, it was quite well known to both the Greeks and Romans and has been fished there for thousands of years. The very earliest illustrations of billfish were undoubtedly swordfish, although these were often rather fanciful, presumably because the early artists rarely had access to fresh specimens as models. Today, the swordfish is widely sought by commercial fisheries around the world, and for at least a century, has been the holy grail of sportfish anglers.

While the broadbill swordfish (Family Xiphidae) is certainly a 'billfish', it is not particularly closely related to the other billfishes – the marlins, sailfish and spearfishes (Family Istiophoridae). The two certainly resemble each other in body form, but closer examination reveals that this resemblance is quite superficial. The most obvious difference is that the bill of the broadbill is flat, while the bills of marlin, sailfish and spearfish are round. Other features of the broadbill which separate it from the istiophorids are its possession of a single, broad keel on the caudal peduncle, a short-based, non-retractable first dorsal fin and the complete absence of pelvic fins.

The broadbill swordfish is one of the most widespread fishes in the world. It occurs throughout the three major oceans, as well as in the Mediterranean, extending from 50°S in some areas to as far as 60°N in the eastern Atlantic. It is caught mainly by high-seas longliners throughout its range, although it is known that the species congregates at seamounts, and tends to move near continents during summer spawning activity.

The only region where significant numbers of broadbill swordfish have been tagged and released is off the eastern United States. Results indicate that the species is capable of extensive movements throughout its range in the Atlantic – distances over 1000 nautical miles (1800 km) being regularly recorded. The furthest distance moved by a tagged swordfish was 2700 nautical miles (5000 km) by a fish tagged off the eastern United States and recaptured off Spain only 390 days later.

Recently, swordfish have been tagged using popup satellite tags (PSATs) in a number of regions, including both the eastern and western United States, the western Indian Ocean, off northern New Zealand, and most recently, off eastern Australia. In one of the first such studies, 29 swordfish caught by longline were tagged off the southeastern United States. Results showed strong northeasterly directed movements over periods of 30 to 90 days. This study was of special interest since most of the swordfish tagged were very small but still showed a strong tendency to travel. For example, one little swordfish, weighing only about 4 kg (9 lb), moved 3050 km in 90 days.

A second study off southern California tried to avoid hook-related mortality by using expert harpooners to place PSAT tags in 16 free-swimming swordfish. Only five of the tags successfully reported, and these indicated considerable movement out into the wider Pacific, some reaching about two-thirds of the way to Hawaii in about five months.

Because swordfish are so widespread, and because only one stock is thought to exist in the Atlantic, it has often been assumed in the past that there is a single stock in the Pacific. However, recent genetic studies are now strongly suggesting that there are almost certainly separate stocks of swordfish, at least in the western and eastern Pacific.

A number of studies have examined growth rates of broadbill in different parts of the world. The results of these vary considerably, some suggesting slow growth, others indicating rapid growth, at least in the early years. Of the more reliable studies, most agree that fish reach about 120 cm (lower jaw to tail fork), or about 20 kg (44 lb) by one year old. By the age of five, fish have probably reached about 100 kg (220 lb), and by ten years of age, sizes of about 150 kg (330 lb) would be attained. Current work will no doubt refine these estimates. As is the case with some of the istiophorid billfishes, male and female swordfish grow at similar rates for about the first three years of life, but then diverge, with females growing faster, and ultimately, living longer than males.

As swordfish grow through the juvenile phase, they undergo a type of metamorphosis in body form; that is, they change rather dramatically in appearance. Up to a size of about 3 to 4 kg (7 to 9 lb), their bodies are covered with rows of sharp tooth-like spikes, the lower jaw is very long, and both jaws contain numerous small teeth. Gradually, the lower jaw stops growing, the spikes recede into the skin and the teeth disappear completely, the swordfish eventually taking on the adult appearance by the time they reach about 10 kg (22 lb) in size.

The maximum size to which broadbill grow is difficult to determine with confidence. The most oft-quoted maximum is, in fact, the all-tackle world record which was set off Chile in 1953. That fish weighed 536.15 kg (1182 lb) and was considerably heavier than the second heaviest fish which appears in the records of IGFA, a fish of 369 kg (812 lb) caught off New Zealand in 2003. On the other hand, scientific sources sometimes quote a maximum size of 650 kg (1430 lb), but this is not able to be verified.

Larval broadbill have been found over extensive areas in all three major oceans, usually in water with surface temperatures above 24°C. Spawning in the Pacific apparently takes place from March to July in the central region, September to December in the southwest and throughout the year in tropical waters.

A recent study of swordfish reproduction off eastern Australia indicated that active female swordfish spawn eggs in batches averaging 1.66 million as often as every day or two. Another study off Mexico suggested that individual swordfish probably produce hundreds of millions of eggs over their protracted spawning period.

Broadbill swordfish feed over extensive depth ranges, down to at least 700 metres, and therefore eat a very wide variety of food items. In deep water, they feed not only on squid, but on fish species including snake mackerels, lanternfishes and Ray's bream. In the full water column, broadbill feed on a very wide variety of prey items, especially squid, but including crustaceans, many species of fish, and even small sharks. Swordfish almost certainly use their sword to slash laterally through schools of fish or squid, since dismembered and slashed prey items are often found in broadbill stomachs.

One aspect of the behavior of broadbill differs from area to area. In some parts of the world, notably off southern California, and in the Mediterranean, the species is traditionally taken during the day by harpoon. In these regions, it is not unusual to see broadbill on the surface (spotter planes are used in California), but it is notoriously difficult to entice these fish into taking a bait, even if trolled right past their bills. In other areas, how-

J. F. Henn

Juvenile swordfish less than a year old. Even at this size, swordfish are prone to capture on longline hooks.
Julian Pepperell

0   10   20   30   40   50
cm

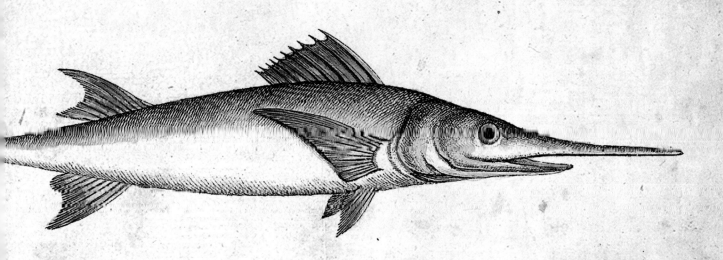

ever, such as off Florida, and the Australian east coast, it is very rare to encounter a broadbill at the surface. Why is this so? The accepted theory is that, where it does happen, fish come to the surface after feeding in very cold, deep water to bask, and warm up, thereby aiding digestion. Presumably, in areas where broadbill do not bask on the surface, either the temperature is not so cold at similar depths, or the fish do not need to make such deep dives to find food.

Some detailed knowledge of vertical movements and depths of diving has been revealed by electronic tracking. The first broadbill to be tracked in this way were tagged off Baja California in 1977. These studies, among the first tracking studies to be undertaken, were pioneered by the late Frank Carey of the Woods Hole Oceanographic Institute. The results of this early work were dramatic indeed, clearly showing that broadbill regularly dived to great depths. They also showed for the first time the daily (diurnal) pattern of diving to depths during the day, and ascending to near the surface during the night. One likely reason for this type of behavior appears to be related to feeding. It is well known that many smaller organisms in the open ocean, including squid, lanternfishes, crustaceans and larvae, undertake daily vertical migrations, descending at dawn and ascending at dusk.

From Carey's early work, it was obvious that swordfish could tolerate much colder temperatures than most other pelagic fish. Again, sonic tracking provided evidence of this, and led to some further intriguing discoveries. Muscle probes in tracked fish showed that they could maintain their body temperatures some 3–5°C above that of the surrounding water while probes placed near the brains also revealed that this organ was being kept at up to 10°C warmer than surrounding water. It is now known through the work of Barbara Block, now

Above An 18th-century illustration of a swordfish. These early depictions › were not accurate, but nevertheless have a great deal of charm.

Below Illustration of a postlarval swordfish measuring 46.3 mm from the tip of the lower jaw to the base of the tail. Note the two jaws of nearly equal length, the tiny teeth in both jaws and the body and head covered with small denticles (spikes). *From Richards, WJ (1989), NOAA Technical Memorandum NMFS-SEFC 240.*

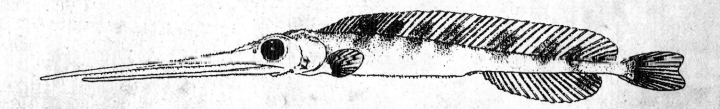

at Stanford University, that both the brain and eye are kept warm by a modified eye muscle. The muscle can no longer contract, but has been converted into a heat-generating organ, positioned so that the brain and retina are constantly warmed. In this way, the predatory swordfish can remain more alert and can see more efficiently than its cold-blooded prey.

The broadbill swordfish has long been an important commercial species, and is a highly regarded food fish, especially in Europe and the United States. While broadbill has been primarily a bycatch of the oceangoing longline fleets of the world, it is increasingly being targeted to supply growing demand.

Total world catches of swordfish are of the order of 80 000 tonnes per annum. Of that total, about 30 000 tonnes are taken in the Pacific and at present, biologists consider the broadbill stocks of the Pacific to be in relatively good condition. Most of the Pacific catch was in the eastern Pacific until a US longline fishery for broadbill started in Hawaii in the mid-1980s. That fishery expanded considerably, fishing seamounts further and further away from Hawaii. Also, off southeastern Australia, a longline fishery targeting

broadbill expanded very rapidly. From an annual catch of a few tens of tonnes in the mid-1990s, the catch from a relatively small area quickly increased to over 2000 tonnes. Subsequent declining catch rates strongly indicated localized depletion of that stock. A quota was put in place and catch rates recovered somewhat within three years.

In the Atlantic, where swordfish have been heavily fished, it is well documented that stocks have been overfished to a level well below the sustainable yield for some time. Considerable international efforts have been made to arrest this situation, and there are some signs that recruitment is improving in this fishery. Fortunately, the Pacific Ocean is a big place in comparison, and prior knowledge of what has happened in the Atlantic may help to avoid a similar scenario from occurring there. For many sportfish anglers, the broadbill swordfish really is the holy grail, even though most will never see, let alone catch, one in their lifetime. Even so, this unique fish holds a very special place in the hearts and minds of anglers around the world. Simply knowing that they are out there is important, and to many, that is enough.

6 Tunas

TUNAS

YELLOWFIN TUNA
Thunnus albacares

OTHER COMMON NAMES

	Allison tuna
	Fin
France	Albacore
Japan	Kihada
Spain	Atún de aleta amarilla
	Rabil

By any standards, the yellowfin tuna is an impressive fish. Beautifully adapted and streamlined for its oceanic existence, it has been given many names, but none more evocative than the Hawaiian name 'Ahi', meaning 'Fire'. One can imagine those early, intrepid Polynesian fishermen yelling 'Ahi!' as their cord lines burnt their calloused hands with the blistering first runs of those keenly sought tuna.

In 1905, game fishing pioneer Charles Holder and his compatriots were surprised by the appearance of a 'new' type of tuna off southern California – a 'yellow finned' tuna. Over 500 of these fish, up to 27 kg (60 lb) in size, were caught on rod and reel that season – almost certainly the result of an El Niño event which had brought the yellowfin north from their more usual Mexican habitat. Holder speculated that such fish might well grow to 100 lb (45 kg). We now know that the yellowfin attains more than three times that size and is one of the most prolific tunas in the world.

Above a size of about 35 kg (77 lb), yellowfin tuna are easily identified by their extended anal and second dorsal fins, which become extremely long in large adults. Fish less than 15 kg (33 lb) may be confused with other species, espe-cially the bigeye tuna. Yellowfin and big-eye tunas can be separated by examining the surface of the liver, which is marked with dark streaks (striations) in the big-eye, but is plain colored in the yellowfin. Externally, the belly and flanks of live yellowfin show numerous vertical sil-very rows of spots, alternating with solid bands, while in the bigeye, these bands are widely separated and mostly solid.

The yellowfin tuna has a worldwide distribution which includes the Atlantic, Indian and Pacific oceans. While yellow-fin tuna are generally classified among the 'tropical' tunas, they sometimes frequent much cooler water. They certainly spawn in the tropics, between latitudes 15°N and 15°S, but their geographic range is much broader – usually between latitudes 40°N to 40°S.

Although genetic studies have shown that the yellowfin is a single, worldwide species, it is very likely that localized, regional populations exist. The body shapes of fish can vary from place to place (as determined by computer analy-sis), probably reflecting transient 'races' of fish. For example, off southeastern Australia, yellowfin tuna encountered in offshore waters were found to have longer anal and second dorsal fins than

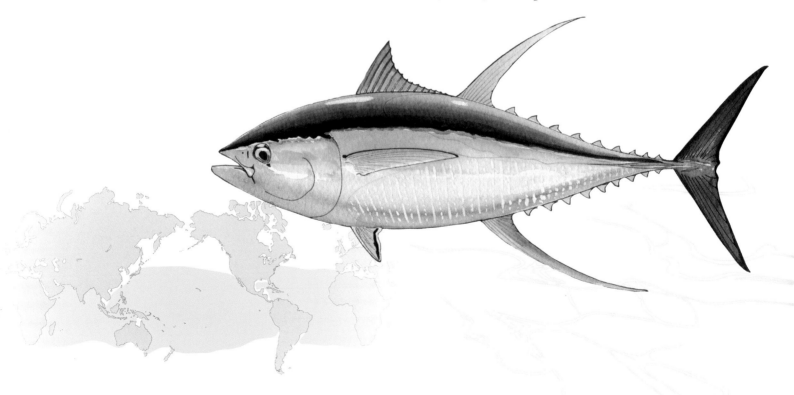

On the move: a hunting school of yellowfin tuna in tight formation
boyceimage.com

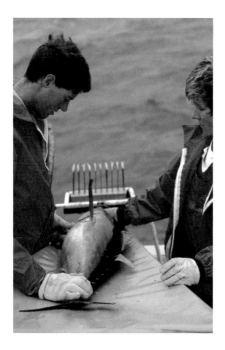

inshore fish of the same size. Why this should be so is not clearly understood, but it probably indicates that the inshore fish live quite separately from the oceanic fish for at least their first two years of life.

Like other pelagic fish, movements of yellowfin tuna have been studied primarily by means of tagging. Scientific tagging studies have been critical in determining not only movements, but stock structure and exploitation rates of the species throughout its range.

Significant numbers of yellowfin have also been tagged by recreational anglers operating through the Australian Gamefish Tagging Program. Since the inception of the program, just over 35 000 yellowfin have been tagged and about 670 recaptures have been reported. While a few tagged yellowfin have made extensive movements, these tend to be exceptions to the rule. In fact, results indicated little movement away from the continental shelf for up to two years, but rather, a consistent seasonal movement of fish north and south along the coast in response to sea surface temperature. These results strongly suggest that yellowfin which find their way onto the continental shelf tend to remain as a group within the coastal strip, effectively

separated from those encountered in the wider western Pacific.

Another major tuna tagging operation was undertaken off Hawaii in the early 2000s. There, yellowfin and bigeye tuna were tagged over several seamounts where there is targeted commercial pole-and-line fishing for tuna. Recapture rates of tagged fish average about 10 per cent, from fish recaptured not only at the seamounts, but also near the island of Oahu, indicating frequent exchange between the island and the seamounts. As part of this exercise, yellowfin tuna were also tagged from sport fishing boats at Midway Island, at the northwestern tip of the Hawaiian chain. Some recaptures of those fish showed quite extensive movements to the west, proving that at least some yellowfin may make long movements across open ocean. However, the emerging picture from all this tagging is that the average yellowfin tuna does not undertake long-distance movements during its lifetime.

Yellowfin tuna grow fast, perhaps faster than any other tuna. From the size of a pinhead at hatching, they reach about 15 kg (33 lb) after only one year, 30 kg (66 lb) plus by year two, and by their fifth birthday, the average fish would weigh about 55 kg (120 lb), while a 70 kg (155 lb) fish would be no more than seven years old.

Some of the largest yellowfin in the world have been caught near the Revillagigedo Islands, about 400 km west of Cabo San Lucas, Mexico. Long-range

Above left
Scientific tagging of small yellowfin tuna has been critical in defining their movements.
John Diplock

Below A baby yellowfin tuna, measuring about 20 cm long, from the northern coast of Papua New Guinea. This fish would be in its first few months of life.
Julian Pepperell

sport fishing boats make regular trips to these grounds, and many of the heavy tackle world records have been, and are held by fish from this area. The all-tackle game fishing record for yellowfin tuna was caught in 1977 near these islands, and weighed 176.3 kg (388 lb). However, the dominance of these grounds has been recently challenged by the discovery of a new hotspot for giant yellowfin – in the eastern Atlantic off Ghana. In 2003, a 175 kg (385 lb) fish was caught in that region, followed in the same year by the all-tackle women's record yellowfin of 142.4 kg (314 lb).

Even though some references suggest that yellowfin might mature at a size as small as 60 cm in length – only about 4 kg (9 lb) – it is much more widely accepted that most yellowfin mature at 25 to 30 kg (55 to 66 lb) in weight, or about 100 cm long. Full maturity is reached by about 45 kg (100 lb), or three years old. Most spawning occurs in the tropics between the latitudes of 15°N and 15°S in the three major oceans.

A remarkable feature of yellowfin reproduction (and probably that of all other tunas) is that a female fish may spawn every day or every second day for several months. Each time she spawns, she will produce several hundred thousand eggs; so over her lifetime, she will have

cast many tens of millions of potential offspring to the mercies of the currents.

Fertilization is external, and mortality of eggs and larvae must obviously be very high. The larvae have large eyes and mouths, and immediately begin actively swimming and pursuing food in the form of other plankton. It is at this critical stage in the life cycle of the yellowfin when slight increases or decreases in mortality will determine how good or bad fishing will be when this 'year class' of fish grow to a fishable size (this is termed 'recruitment' of fish into a fishery). It has been found that good years and bad years for catches are closely linked to the strength of these recruitment events.

Adult yellowfin tuna are opportunistic predators, meaning that they feed on quite a wide variety of prey items, including fish, squid and crustaceans. They possess comb-like gill rakers which can strain largish food particles from the water, which helps to explain why their stomachs often contain quite small fishes, squids and postlarval oceanic crabs.

Daily patterns of behavior of yellowfin tuna have been extensively studied using various electronic tags. This research has revealed that yellowfin tuna are surface-oriented fish, preferring to remain above the thermocline most of the time (the thermocline being the depth

where warmer surface waters meet colder, deeper water, usually less than 100 metres below the surface). Data from many fish also showed that they are even more surface-associated at night compared with daylight hours. One interesting finding from tracking yellowfin in Hawaii was that fish associated with surface buoys or islands will move in a highly predictable daily feeding pattern. Fish tagged near a particular fish aggregating device (FAD) would always leave the FAD at the same time, usually around dusk, and swim in a consistent circuit to other FADs, often many kilometres away, completing their trip by dawn the next day. While these results suggest that yellowfin may be creatures of habit, it was also shown that the repeat behavior patterns don't go on indefinitely. After periods of a few days to a week or so, fish would depart the scene, presumably to resume their more nomadic lifestyle.

In recent years, larger numbers of yellowfin have been implanted with archival tags, revealing even more remarkable details of their behavior. Again, fish stayed nearer the surface at night compared with daylight hours, making forays to around 150 metres below the surface

every now and then. However, the big surprise from this study was that some fish made far deeper dives than ever before recorded, some exceeding 1000 metres below the surface. As well, deep-diving fish often remained at these depths, where temperatures were less than 5°C, for up to an hour, proving that these remarkable fish have broader tolerances than previously thought.

The yellowfin tuna is an enormously important commercial species. The total world catch is at least 800 000 tonnes, over half of which is taken in the central and western Pacific Ocean. For many years, the global catch and western Pacific Ocean catch of yellowfin continued to rise, but in recent years, the catch has shown signs of flattening out. Changes which have occurred include a reduced total catch of yellowfin by longline and a greatly increased catch of juvenile fish by purse seine and by artisanal fishermen in the Philippines and Indonesia. Sophisticated stock assessments of yellowfin and other tunas are carried out each year and have begun to sound some warning bells about the viability of continued high exploitation rates of this species.

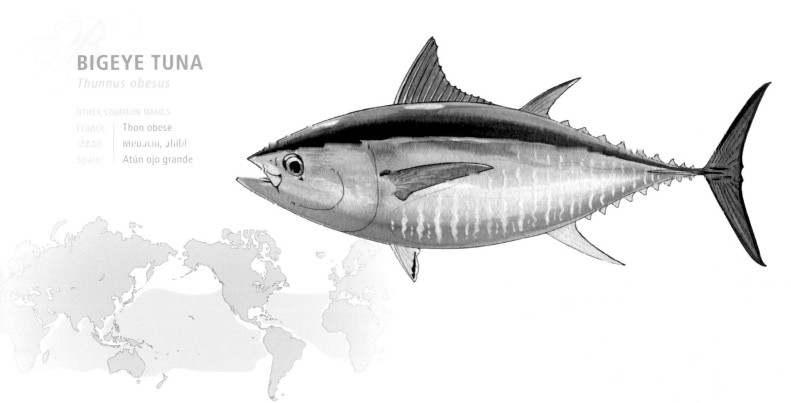

BIGEYE TUNA
Thunnus obesus

OTHER COMMON NAMES
France | Thon obese
Japan | Mebachi, shibi
Spain | Atún ojo grande

The scientific name of the bigeye, *Thunnus obesus*, simply – and perhaps unflatteringly – means 'fat tuna'. And compared with other tuna species, this is indeed an apt description. This is a large, deep-water tuna species which is highly sought by international fisheries around the world.

The species with which the bigeye tuna is most often confused, especially at smaller sizes, is the yellowfin tuna. It is statistically true that the bigeye does have a slightly larger eye than the yellowfin, but only on average, so this is not regarded as an absolute or diagnostic character to separate the two. However, there are a number of features which are diagnostic at all sizes. Internally, the liver of the bigeye has three lobes of roughly equal length, the lower surface of which is streaked with many fine, blood-colored lines, especially at the edge of the middle lobe. The liver of yellowfin has one lobe longer than the other two, and is unstreaked. Externally, the tail of the yellowfin tuna has a yellow to golden tinge while the midpoint of the trailing edge of the tail is indented into a distinct 'V' with two raised ridges on either side. In comparison, the tail of bigeye tuna is dark, showing little if any yellow coloration, while it shows no distinct 'V' indentation nor a pair of raised ridges.

Bigeye tuna have a very broad distribution, extending between about 40°N and 40°S in all three major oceans, except in the southwestern sector of the Atlantic. The depth of the thermocline is a critical factor in the distribution of the bigeye, partly because of temperature. The thermocline marks the boundary between the mixed, warm surface layers of the ocean, and the deeper, colder water masses. Many surface fishes, such as the marlins and yellowfin and skipjack tuna, tend to treat the thermocline as the 'bottom' of their environment, whereas bigeye tuna often treat it as an upper boundary. The thermal preference for bigeye tuna is between 17°C and 22°C, which is about the temperature range of the permanent thermocline throughout the species' range. Bigeye tuna prefer to stay near, and usually below the thermocline, but come to the surface periodically. One of the most fascinating aspects of the distribution of bigeye is related to its ability to dive to, and remain at considerable depths. This is the ability, at least of adult fish, to tolerate lower oxygen levels than any of the other tropical tunas, and at temperature ranges of only 11–15°C. So, even though bigeye tuna do have a tropical to subtropical distribution, they are far more tolerant of habitat extremes than their close relative, the yellowfin tuna.

Virtually nothing was known about the movements of bigeye tuna in the western Pacific until Australian scientists fortuitously tagged several hundred off Cairns in the late 1980s (they were targeting yellowfin tuna at the time). Much to everyone's surprise, several tagged bigeye tuna were recaptured in subsequent months, and years, thousands of kilometres to the east, in contrast to results from tagging of yellowfin tuna at the same time and place.

On the other hand, a major tuna tagging program which has more recently been undertaken off Hawaii has produced somewhat contrasting results. Results of that program, undertaken over seamounts off Hawaii, indicate that juvenile yellowfin tuna tended to move relatively short distances, mainly between seamounts and the main Hawaiian islands, while bigeye tuna moved more randomly, but not to any great distances. In fact, the furthest distances moved on this study have been for yellowfin, but this may simply be because those have been large fish.

Because bigeye tuna tend to spend a lot of their time in relatively cool water,

it was suspected that their growth rates would be slower than the more surface-oriented yellowfin. And a recent definitive study by scientists from CSIRO and other agencies has confirmed that theory. By counting what are assumed to be daily growth rings on otoliths (ear bones) of bigeye and then annual growth rings, it has been estimated that fish average one year old at a fork length of about 80 cm (about 9 kg – 20 lb), just under 100 cm (18 kg – 40 lb) at two, 115 to 120 cm (28 kg – 62 lb) at three and about 130 to 135 cm (45 kg – 100 lb) at five years old. In comparison, a two-year-old yellowfin would be about 25 kg (55 lb) and a five-year-old, about 60 kg (130 lb). Bearing in mind that these estimates are averages and may vary from place to place, it nevertheless indicates that bigeye tuna do indeed grow more slowly than yellowfin. As well, the researchers found that the oldest bigeye tuna were 16 years old – compared with yellowfin, very few of which would attain ten years of age.

Even though bigeye tuna grow more slowly than yellowfin, their longer life span means that they attain a greater maximum size in most areas. For example, the all-tackle world record Pacific

The contrasting body shapes of a yellowfin (top) and a bigeye tuna (bottom)
Julian Pepperell

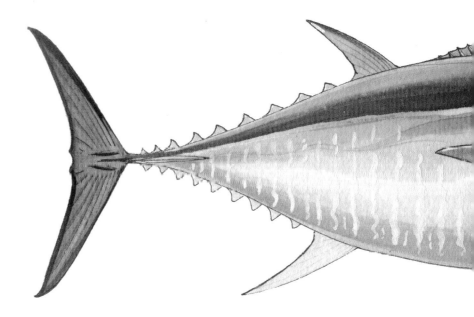

bigeye tuna weighed 197 kg (435 lb), while in the Atlantic, the record is 178 kg (392 lb). In contrast, very few yellowfin tuna over 150 kg (330 lb) have been caught, with the largest, weighing 176 kg (388 lb), caught off Mexico. These very large bigeye tuna must be considerably older than the largest yellowfin, and while this longevity may also be related to energy and metabolism, the need for research into the different growth rates of the two species is apparent.

Spawning of bigeye tuna occurs throughout tropical regions of all three oceans. Mature, spawning fish have been identified from south of Indonesia, in the Coral Sea and north of Australia and in the central and eastern Pacific between 15°N and 15°S. As is the case with many tropical pelagic fishes, bigeye tuna spawn mainly in late spring and summer months in the respective hemispheres, but spawning has been observed in most months if surface temperatures exceed about 24°C.

Bigeye tuna mature during their third, or sometimes fourth year of life, at a size of between 100 and 130 cm fork length (20 to 40 kg – 44 to 88 lb). This is a similar size for spawning, but at an older age compared with yellowfin, which usually mature in their second year of life at about the same size. Again, this is indicative of a fundamental physiological difference between the two species.

The larger the bigeye tuna, the more likely it is to be a male. This phenomenon is also similar for yellowfin tuna, but is the opposite for marlins. The reason for the difference in bigeye and yellowfin is usually surmised to be because the females put so much energy into producing eggs that there is little left over for growth.

Although it had been known for many years that bigeye tuna tended to swim at depth (because they were caught on the deepest longline hooks), and that the thermocline was a key feature of their environment, it was not until they were tagged with electronic devices that the true nature of their vertical swimming behavior began to be realized.

Early experiments showed that bigeye tuna generally stay deeper than yellowfin, and are inclined to make occasional, very deep dives. After being at depth for a while, fish tend to make regular rapid ascents in the water column, returning to the previous depth. A particularly interesting aspect of this behavior was that the fish were ascending very quickly but then descending much more slowly. Almost certainly, this was proof of a form of thermoregulation whereby bigeye would warm themselves at shallower depths and lose heat more gradually as they searched for food in deeper, colder water.

Bigeye tuna have amazed physiologists with their abilities to dive rapidly to great depths and to withstand quite low oxygen concentrations in doing so. Recent studies have shown that depths attained often exceed 700 metres, and on occasion, have reached 1700 metres below the surface. So low are the oxygen levels at such depths that it was surmised that bigeye must undertake the fish equivalent of 'holding one's breath' during these dives. However, it is now apparent that the blood of bigeye tuna is ultra-efficient at transporting dissolved oxygen – another remarkable adaptation of this extraordinary fish.

Over the past two decades or so, the bigeye tuna has been increasingly sought by deep-setting longline vessels since it commands a high price on the Japanese sashimi markets. In the wider Pacific, bigeye tuna only accounts for about 7 per cent of the total tuna catch by weight (a figure which includes skipjack tuna), but even so, the 2007 total catch in the Pacific was 190 000 tonnes, the total value of which far exceeded US$1 billion. Changes in fishing practices and increased catches of juvenile bigeye tuna have caused a degree of concern among scientists regarding the status of bigeye stocks around the world. The main shift has been towards the use of artificial fish aggregating devices by purse seine fleets, and as a result, the proportion of adult fish in the total catch has declined considerably, even though the total catch has been maintained or even increased. There is also some concern over apparent declining catch per unit effort of bigeye, and the species is now regarded as being overfished, at least in the central and western Pacific, and possibly in the Indian and Atlantic oceans as well.

ALBACORE
Thunnus alalunga

The albacore was one of the earliest tunas to be harvested for human consumption. Its non-tropical surface distribution, especially in the Mediterranean, its relatively small size and its schooling habit made it ideal for early European fishermen to catch. The ancient word 'albacore' simply means 'tuna' in Portuguese and Spanish, although the origin of the name is quite obscure. Some assume that it comes from the Latin 'alba' meaning 'white' and 'cora' meaning 'flesh'. However, others dispute this, insisting that the name derives from the Arabic 'al-bacora' – 'the little pig' (presumably because of its fatness, and also its white meat). Its scientific name, *Thunnus alalunga*, simply means long-winged tuna – an appropriate description of this graceful fish.

Because of its name, the albacore is not always thought of as a tuna. However, it is indeed a fully fledged member of the 'true' tunas, classified in the genus *Thunnus* along with the yellowfin, bigeye, southern and northern bluefins, longtail and Atlantic blackfin tuna. The albacore is easily identified by a number of clear, diagnostic features. The trailing margin of the tail is white, the only tuna in the genus *Thunnus* to exhibit this feature. At all sizes, the pectoral fins are extremely long and strap-like (half sickle-shaped when viewed from above), always reaching beyond the level of the origin of the anal fin, often further. The two dorsal fins and the anal fin are sometimes, but not always, tinged with yellow. The surface of the liver of albacore is streaked with dark blood lines, unlike that of the yellowfin, which is uniformly plain colored. Lastly, the depth of the body of the albacore is greatest at the midlevel of the first dorsal fin.

The albacore is a true 'cosmopolitan' species, having a worldwide distribution. It occurs throughout the Atlantic Ocean as well as the Indo-Pacific and because of its relatively broad temperature tolerance, extends to the rather high latitudes of 50°N and 50°S in the three major oceans. On an oceanic scale, the distribution of albacore expands towards the poles in summer in each hemisphere, and contracts towards the equator in winter. Albacore are an important catch of longline fisheries, especially those that set their hooks to fish below the thermocline in order to target the lucrative bigeye tuna. In so doing, catches of albacore have increased considerably, especially in the western Pacific. Thus, when a map of the worldwide distribution of albacore is considered, it is perhaps surprising to find that the species is commonly caught not only in the cooler latitudes, but well into the tropics. However, as noted, all of these subtropical fish are caught by longline well below the surface where the cooler habitat is more suitable. Through out their range, whether in the tropics or higher latitudes, albacore tend to prefer temperatures in the range 16–22°C, but have been caught in water as cold as 2°C. They make excursionary dives to depths of at least 500 metres, and probably considerably deeper.

As early as the late 1800s, it was speculated that the albacore which regularly appeared and disappeared off the Californian coast must be migratory by nature. This was well before tagging of tunas was made possible by the development of the plastic dart tag, but it seemed clear to the fishermen of the day that the albacore must move with distinct water masses on a seasonal basis. As we now know from tagging studies, these ideas were eventually to be proven correct.

Relatively few albacore have been tagged compared with species such as

yellowfin and skipjack tuna, but even so, the results have been very interesting. In the northern hemisphere, albacore have mainly been tagged from commercial trolling vessels – relatively small dories which troll an array of simple lures over remarkably long distances. This has shown that in both the Pacific and Atlantic oceans, albacore are capable of making transoceanic crossings, the first species shown to do this in the Pacific. Recorded long-distance movements have been from west to east, and vice versa in both oceans, with coastal north–south movements as well. However, there has never been a recapture of an albacore in one hemisphere which had been previously tagged in the opposite hemisphere. In other words, as noted above, albacore populations in the northern and southern Pacific and Atlantic are quite distinct from each other, and therefore, are managed as different stocks.

Off eastern Australia, over 17 000 albacore have been tagged by recreational anglers, and 157 recaptured, a similar recapture rate to that of commercial tagging programs in the western Pacific. One interesting result from this tagging activity is the limited average movement exhibited by recaptured albacore. Most recaptured fish moved less than 400 km, even after periods of liberty of two, three and even five years.

The growth rate of albacore varies considerably depending on geography. By studying growth via a number of methods, including scale and otolith (ear bone) analysis and the progression of fish sizes through time, a range of possible growth rates have been proposed for different areas around the world. The latest work in the south Pacific suggests that albacore attain a length of around 45 to 50 cm by their first birthday, but slow down to reach 60 cm by three to four years old. Longevity is about 19 years, but in reality, very few fish exceed ten years old. This growth rate is quite slow compared with other tunas such as yellowfin, and has been confirmed by recaptures of tagged albacore. As just one example, a very small albacore estimated at about 2 kg (4.5 lb) in weight which

The extremely long, curved wing-like pectoral fins of the albacore
boyceimage.com

I tagged in the southwestern Pacific was recaptured five years later weighing 13 kg (29 lb).

The all-tackle game fishing record for albacore is 40 kg (88 lb). That fish, and a number of other large albacore, were caught near the Canary Islands in the Atlantic Ocean. Other capture locations of albacore weighing in excess of 30 kg (66 lb) include Catalina Island, San Diego and San Pedro (all off southern California), Cape Point, South Africa and Miyake Island, Japan.

The maximum weight for albacore cited in the literature is 60.3 kg (133 lb). It is not clear if this is an actual weight, or derived from a length measurement, but either way, if correct, this was truly a giant albacore.

Albacore spawn in distinct bands in subtropical latitudes (latitudes of 10° to 25°) in both hemispheres in the Pacific and the Atlantic oceans, and in a single band across the southern Indian Ocean. This marked separation of spawning areas reflects an actual separation of stocks of albacore north and south of the equator, which has been verified by many studies.

Studies suggest that albacore spawn in tropical and subtropical waters (latitudes of 10° to 25°) throughout summer, each female producing two to three million eggs over this period. It is interesting that, like yellowfin tuna, male albacore grow larger than females, even though females mature at a larger size (about 85 cm fork length) compared with males (about 60 cm). It is thought that this dif-

ferential is due to the high energy costs for females in constantly producing eggs during a spawning season, at the expense of growth.

Albacore are caught commercially by two main methods, trolling (by American, New Zealand and Canadian boats) and longlining (mainly by Taiwan and Korea). The catch of albacore in the southwestern Pacific (remembering that the northern Pacific stock is completely separate) has been just under 100 000 tonnes since the mid-1990s. Longlining accounts for about 60 per cent of this catch. As an indicator of how fisheries can rapidly change, in the late 1990s, a fleet of hundreds of relatively small longlining vessels expanded rapidly in island nations such as Samoa, New Caledonia, Vanuatu, French Polynesia and Fiji. These small-scale fisheries specifically target large albacore below the thermocline, and their catch grew enormously to reach about 30 000 tonnes by the early 2000s. While the stock of albacore in the southern Pacific is generally regarded as being sustainably fished, this sudden increase in the taking of large, presumably spawning fish has probably resulted in localized depletions, and is being closely monitored. A troll fishery across the southern Pacific targets much smaller fish than longliners. The catch in that fishery has fluctuated quite widely, probably dependent to a large extent on the ability of the vessels to find concentrations of fish. Nevertheless, the catch of that fishery is now relatively low, at around 4000 tonnes in 2005.

ATLANTIC BLUEFIN TUNA

Thunnus thynnus

OTHER COMMON NAMES

	Northern bluefin tuna
	Giant bluefin tuna
	Tunny
France	Thon rouge
Japan	Kuro maguro, Hon maguro
Spain	Atún

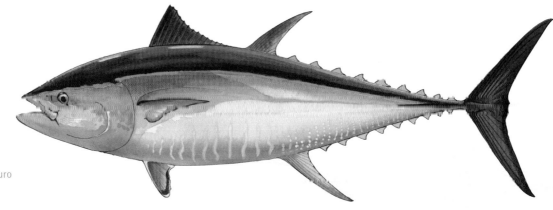

The Atlantic bluefin tuna holds a special place in the history of human contact with, and fishing for open-ocean fishes. The species has been consistently caught in Mediterranean set nets for thousands of years, while along the eastern seaboard of Canada and the northern United States, giant bluefin exceeding 454 kg (1000 lb) in weight were being harpooned for their oil in the mid-1800s. Until recently, the identity of the bluefin tunas was a little uncertain. While three bluefin tunas are recognized, their status as species or subspecies was unclear. It has now been determined that each is a true species in its own right, as follows: Atlantic bluefin tuna, *Thunnus thynnus*, northern Pacific bluefin tuna, *Thunnus orientalis*, and the southern bluefin tuna, *Thunnus maccoyii*.

A number of features may be used to separate the Atlantic bluefin tuna from other Atlantic tunas. It is a plain colored tuna without the yellow horizontal flash of the yellowfin tuna. Its pectoral fins are short, usually not longer than 80 per cent of the head length. It has a striated liver (the liver of yellowfin tuna is plain colored) and it has 31 to 43 gill rakers on the first gill arch (blackfin tuna have

19 to 28 and bigeye tuna have 23 to 31). Of course, the bluefin grows to a much larger size than the other species, so any Atlantic tuna bigger than 200 kg (440 lb) would automatically be a bluefin.

The Atlantic bluefin tuna occurs throughout the northern Atlantic. Its broad thermal tolerance is reflected in its range, from Newfoundland and Scandinavia in the north to the Canary Islands in the south. Importantly, and as noted, it also occurs throughout the Mediterranean.

Seasonal movements of Atlantic bluefin tuna were known to occur thousands of years ago, at least into and out of the Mediterranean. The great schools of tuna enter the Mediterranean each year and, although the numbers fluctuate over time, the schools take the same route, at the same time each year. The question, of course, was where had they been in the meantime and where did they go? The answers were finally revealed with the advent of tagging. The Atlantic bluefin tuna was the first pelagic species to be tagged in large numbers to determine movements. Innovative researcher Frank Mather III of the Woods Hole Oceano-

graphic Institute pioneered the tagging of bluefin tuna with plastic dart tags, and in the early 1950s, the first transatlantic crossing of a tagged tuna was recorded (from west to east). Even though such movements may have been suspected, the proof that these fish could, and did traverse an ocean basin was a watershed in fisheries thinking. Further tagging confirmed these movements, but there was still a degree of uncertainty over whether there were one or two stocks of bluefin tuna in the Atlantic and Mediterranean. New tools were needed to address this question.

Since the late 1990s, 960 Atlantic bluefin tuna ranging in size from 45 kg (100 lb) to 450 kg (1000 lb) were tagged in the northern Atlantic using both archival and popup satellite tags. This tagging has shown a clear demarcation of western and eastern Atlantic/Mediterranean populations. Results showed that, even though fishes from both populations mix on feeding grounds, they show strong fidelity to their own spawning grounds where they apparently return to spawn every year. One tuna which retained its tag for a long time moved from the

LE THON.   LE MAQUEREAU.

northeastern United States to the Florida Straits spawning grounds three years in a row – a very important finding in relation to international management of tuna stocks. How the two stocks came to be is rather academic, but of considerable interest nevertheless. There is an underwater ridge in the middle of the Atlantic, but that would not appear to be an obstacle for highly mobile pelagic fish. Of course, the emergence of two stocks may have simply been a quirk of evolution due to the chance establishment of a second spawning ground.

There have been many attempts to estimate the growth rates of Atlantic bluefin tuna. These indicate that western Atlantic fish have slower growth rates than eastern Atlantic and that fish mature at an older age in the west than in the east. Compared with the growth rates of tropical tunas, growth of Atlantic bluefin is slow. Fish grow to about 4 kg (9 lb) in their first year, compared with perhaps 15 kg (33 lb) for yellowfin tuna. In the western Atlantic, bluefin tuna first spawn

at about eight years of age, by which time they would weigh about 140 kg (310 lb). In contrast, eastern Atlantic fish start spawning when about five years old, or 60 to 70 kg (155 lb). It is estimated that the maximum age to which Atlantic bluefin tuna live is 20 to 25 years.

The Atlantic bluefin is by far the biggest member of the tuna/mackerel family, Scombridae, and indeed, one of the largest of all the bony fishes. The maximum size recorded for a bluefin tuna is 679 kg (1496 lb) and 304 cm fork length. This massive fish was caught by rod and reel off Nova Scotia, Canada, in 1979. A Cuban scientific paper mentions a fish weighing 684 kg (1505 lb), but the source could not be verified. Many fish have been caught on both sides of the Atlantic in excess of 454 kg (1000 lb).

The Atlantic bluefin tuna spawns in two well-defined areas – at the mouth of the Gulf of Mexico and inside the Mediterranean. The existence of these two widely separated spawning grounds led to the early assumption of distinct west-

Two large Atlantic bluefin tuna showing the characteristic short pectoral fins of the species
Guy Harvey

ern and eastern stocks of Atlantic bluefin and as noted, recent studies using electronic tags have proven the two stock hypothesis. All tunas are highly fecund, producing huge numbers of eggs over the course of a spawning season. In the case of Atlantic bluefin tuna, mature females each produce in excess of 10 million eggs.

Japanese scientists successfully spawned bluefin tuna in captivity as long ago as 1979, but 'closing' the life cycle of captive fish has proven to be a highly elusive goal. In 2003, scientists at Kinki University in Japan announced that they had finally achieved this feat, with the successful spawning of bluefin tuna that were themselves the offspring of captive tuna. This is an exciting development which could have major implications. Perhaps the demand for this highly prized species might be able to be met by mass production in sea pens, thus easing the pressure on wild populations of this heavily fished tuna.

Of all the tuna species, the Atlantic bluefin tuna has been exploited more intensively and over a longer period than any other. As described in chapter 4, an intensive, highly efficient trap fishery for bluefin tuna has existed in the Mediterranean for thousands of years. Fishers from

many coastal countries have taken part in this fishery, including Spain, France, Italy, Sicily, Tunisia and Algeria. In this passive form of fishing, large, heavy nets are staked out near the coast in areas known to be frequented by seasonal migratory schools of tuna. Traveling fish are corralled in the nets which are then hauled by hand to a circle of small boats and the great fish dispatched amid a flurry of foaming water, blood and commotion.

Changes in fish populations are often correlated with commercial catch rates, with 'catch per unit of fishing effort' being used as an indicator of fish abundance. In most fisheries, records of commercial catches may only be available for decades at the most, but in the case of the Mediterranean bluefin tuna fishery, such continuous records extend back in time for up to 400 years. By studying these records from the archives of several countries, scientists discovered that catches of tuna varied greatly through time, with no obvious trend up or down. They even found evidence for long-term cycles with peaks in abundance every 20 years or so, and even longer patterns repeated about every 100 years. These studies revealed that abundance of fish through time was probably heavily reliant on good years for survival of juvenile fish, but the environmental cues that influence survival are not clear. There is no doubt that fishing has had a considerable impact on bluefin tuna populations, but records of this unique fishery remind us that the environment also plays a major role in determining fish abundance.

Nevertheless, stocks of Atlantic bluefin tuna in both the eastern and western Atlantic are now considered to be badly overfished. It is estimated that the western stock has declined by more than 90 per cent since 1970, while analysis of the eastern stock, severely hampered by lack of knowledge of total catches and age composition of the catch, is also considered to be a fraction of biomass needed to ensure long-term sustainability of catches.

An Atlantic bluefin
tuna cutting out
a blue mackerel
from a school

PACIFIC BLUEFIN TUNA

Thunnus orientalis

OTHER COMMON NAMES

	Bluefin tuna
France	Thon bleu du Pacifique
Japan	Kuromaguro
Spain	Atún aleta azul del Pacífico

The Pacific bluefin tuna was for many years lumped with the Atlantic bluefin as the same species, but has recently been recognized as a distinct species.

Pacific bluefin tuna can be separated from all northern Pacific tuna species by their short pectoral fin, plain coloration and high gill-raker count (more than 33). They are separated from southern bluefin tuna by their dark caudal keels (those of southern bluefin are nearly always yellow), but also see the section on southern bluefin tuna regarding this identification feature.

Pacific bluefin tuna occur across the northern Pacific in subtemperate to temperate waters. Individuals, or small pods of large fish, are occasionally caught in the southern hemisphere, but it is likely that this is simply due to vagrant fish wandering away from their northern hemisphere habitats. As is the case for the other two bluefin tunas, the Pacific bluefin also has a well-defined spawning ground, situated between southern Japan and the Philippines.

Tagging has revealed frequent trans-pacific movements between California and Japan. This migratory route has been confirmed with the recapture of a number of fish tagged with archival or 'smart' tags. One particular fish was even shown to have made the round trip from Japan to the United States and back again. During their journeys across the Pacific, electronically tagged fish have consistently stopped over seamounts where they make regular deep dives, probably to feed on squid.

Like its Atlantic cousin, the Pacific bluefin also reaches very large sizes. The largest commercial record appears to be for a fish weighing 555 kg (1220 lb) and measuring nearly 3 metres long. IGFA did not keep separate line class records for this species for many years, so tuna caught off California were lumped with Atlantic bluefin. The current IGFA record for Pacific bluefin tuna stands at 325 kg (716 lb), for a New Zealand fish caught in 2007.

It is clear from long-term catch records of Pacific bluefin tuna that catches have fluctuated markedly through time, albeit around a long-term average of about 20 000 tonnes. That is about the level of the present-day catch and current stock assessments indicate that at least this species of bluefin tuna is not over-fished.

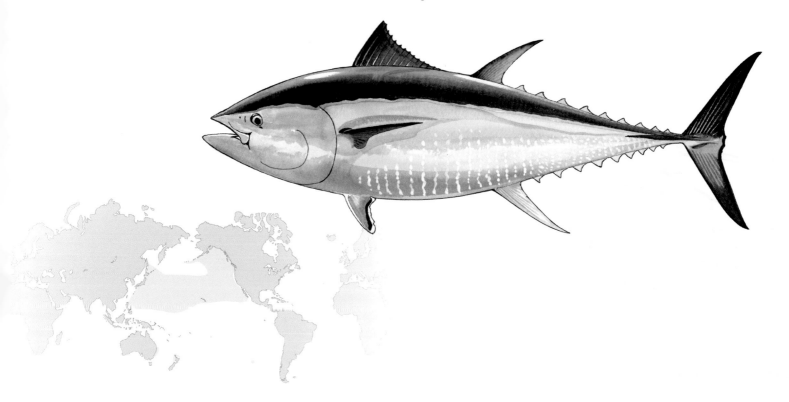

SOUTHERN BLUEFIN TUNA
Thunnus maccoyii

OTHER COMMON NAMES
France | Thon du sud
Japan | Minamimaguro
Spain | Atún del sur

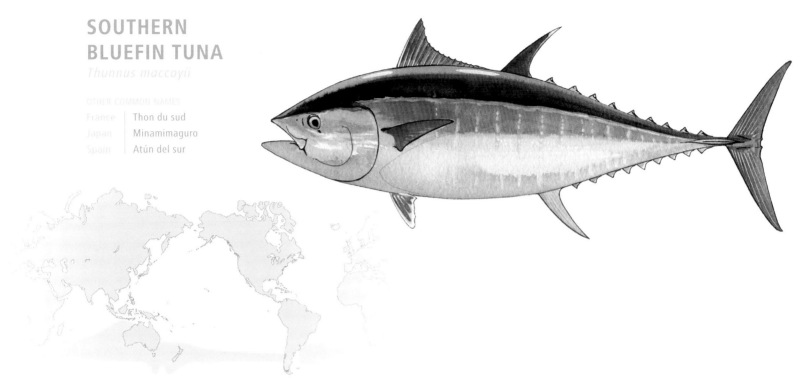

The southern bluefin could be considered a 'classic' tuna in that it is closely related to, and resembles the tuna of the ancients, the bluefin tuna of the Mediterranean Sea and Atlantic Ocean, *Thunnus thynnus*. So similar is the southern bluefin to its Atlantic counterpart, that for many years the two were considered to be subspecies. It is now clear, however, that there are three distinct bluefin tuna species, the Atlantic, Pacific and southern bluefin.

The bluefin tunas are distinguished from most other tunas within their geographic range by their relatively short pectoral fins – usually measuring less than 80 per cent of the head length, and their general lack of body patterning or coloration. The southern bluefin is also a robust tuna compared with the somewhat similar, but more slender longtail tuna, *Thunnus tonggol*.

Occasionally, Pacific bluefin tuna (*Thunnus orientalis*) overlap in their distribution with southern bluefin, especially around New Zealand. To distinguish between Pacific bluefin and southern bluefin, first look at the caudal keels (these are the horizontal projections on either side of the caudal peduncle, or tail wrist). If they are yellow, the fish is a southern bluefin. Sometimes, though,

southern bluefin may have dark keels and even experts can make mistakes in telling the two species apart. There are also difficult internal characters used by taxonomists to split the two bluefins. However, to be absolutely certain, a small muscle sample should be taken and sent to a recognized genetics lab for DNA analysis.

Southern bluefin tuna have an unusual distribution compared with most other tunas. The species is entirely confined to the southern hemisphere, mainly within a relatively narrow band between the latitudes 30°S and 50°S, across the three major oceans. The exception to this overall distribution pattern is the northerly extension of the species in the eastern Indian Ocean, along the northwest shelf of Australia, as far north as Java and Bali in Indonesia. This extension is seasonal and due to the unusual fact that southern bluefin have one, and only one, specific spawning site which is confined to this area (see below).

Regarding thermal preferences, electronic archival tags retrieved from many juvenile southern bluefin reveal that preferred water temperatures are generally between about 7°C and 15°C, but may extend up to 22°C during long-distance movements. As well, during shorter periods of weeks or less, tuna stay within very narrow ranges (thermal niches) with maximum and minimum temperatures varying less than 5°C between the two.

Very young juvenile fish are first encountered moving south along the Western Australian coast. From there, they slowly move into southern Australian waters where one- to five-year-old fish up to about 40 kg (88 lb) – are found while adult fish over about 90 kg (200 lb) occur off southeastern Australia and New Zealand. Intensive tagging experiments on southern bluefin tuna have been conducted in several waves covering many years. Results of tagging many thousands of fish with conventional plastic tags have shown what appears to be a clearcut pattern of movement or migration. Long-distance movements of tagged fish away from the Australian south coast have been recorded numerous times, with most fish moving in a westerly direction across the Indian Ocean towards South Africa. In contrast, movements of tagged fish eastwards to New Zealand and beyond have been far less frequent.

Southern bluefin tuna have also proven to be an excellent species for the pioneering of archival, or 'smart' tags. Because of their known high exploitation rates and, therefore, recapture rates, it has been possible to implant electronic tags into numbers of fish, knowing that at least some of them would be retrieved at later dates. For example, two fish tagged and released off South Australia were recaptured about ten months later quite near their release points. However, the stored data showed that the fish had traveled halfway to Africa and back during this time.

All of the releases of tagged southern bluefin have taken place around the southern half of Australia, and while there have been many thousands of recaptures, some showing transoceanic movements, only a small number have been recaptured on their spawning grounds off Indonesia (or at least reported from there). It also seems odd that, notwithstanding that adult fish are certainly found on the spawning grounds, the size of fish tends to increase with dis-

tance away from those grounds. Thus, very large southern bluefin are found off southeast Australia, New Zealand and South Africa, leading to the question as to whether any or all of these mature fish undertake an annual long-distance trek to the spawning grounds. Perhaps so, but if that were the case, complete absences of large bluefin would occur during the spawning season (September to April) everywhere except the spawning grounds. This is not the case, so perhaps only a proportion of adults migrate to spawn each year. Popup satellite tags have more recently been used on larger, adult fish. Tagging of seven adult tuna off southeastern Australia showed that fish wandered freely around southeastern and southern Australia, with finally one of them making the presumed, but previously unproven journey to the spawning grounds off Java. The question of what proportion of adults might spawn each year still remains, however.

The southern bluefin tuna is among the slowest growing of all tunas, which is a major contributing factor to its having been overfished in the past. The most recent studies on growth rates of the species indicate that, by the time they are one year old, southern bluefin will have reached only about 3.5 kg (8 lb) (55 cm) in size – much smaller than previously thought. (In contrast, a one-year-old yellowfin tuna would weigh 10 to 15 kg – 22 to 33 lb). This relatively slow growth continues, with fish apparently not maturing until they are about 12 years old – about 90 kg (200 lb) in weight and 165 cm long. Again, the yellowfin tuna comparison is interesting: age and size at maturity are about three years old and 40 kg (88 lb) respectively.

The maximum size to which southern bluefin grow is also cause for some conjecture. The largest southern bluefin caught on rod and reel weighed 158 kg (348 lb) and was caught off Whakatane, New Zealand, in 1980. The next two largest – 148 kg (327 lb) and 138 kg (304 lb) – were also caught in New Zealand, while the all-tackle Australian record stands at 125.7 kg (276 lb) for a fish caught off Tasmania in 1985. On the other hand, the scientific literature is fairly vague about the maximum size to which southern bluefin grow. An oft-quoted maximum size for southern bluefin is 'up to 200 cm and 200 kg'; however, since anglers have never caught a fish approaching 200 kg (440 lb), it seems unlikely that southern bluefin actually attain that size.

In contrast to most other tunas which spawn over extensive areas, the southern bluefin has a remarkably discrete spawning ground, confined to a small area off southern Indonesia. That spawning occurs there was proven decades ago by larval sampling surveys which found concentrations of southern bluefin larvae in this area, and only in this area, from September to March, with peak larval catches in January–February. Adult fish with ripe ovaries and testes are also found on these grounds during the same period. Recent studies have shown that the average age of mature fish on the spawning grounds is very old, with most fish exceeding 20 years.

By measuring the temperature inside stomachs of southern bluefin tuna, researchers have found that juvenile fish feed an average of once per day, ingesting up to several kilograms of food each time. Of particular interest was the finding that this feeding event occurred mostly at dawn. Night feeding was relatively uncommon but when it did occur, as might be suspected, it was during the full moon quarter. When migrating over deep oceanic areas, southern bluefin were found to feed much deeper than when near shore, often down to 300 metres below the surface. They also were capable of fasting for five days or more, especially when they were near the northern edge of their migratory range.

The catch history of southern bluefin is not a happy one. Historically, world catches of southern bluefin tuna peaked at about 80 000 tonnes in 1960–61, the bulk being taken by Japanese longliners. During the 1960s, 1970s and 1980s, the total catch declined steeply and steadily, dipping below 20 000 tonnes by the late 1980s. During the late 1980s and mid-1990s, quotas (total catch limits) were steadily imposed on all components of the fishery. Unfortunately, it has since come to light that not all countries agreeing to the quotas maintained their catches at or below those levels, resulting in higher catches than were considered sustainable. Real fears are now held for the long-term recovery of the stocks, which are now thought to be as low as 7–15 per cent of the original parental biomass.

Southern bluefin tuna being transported at sea to a coastal tuna farm. Note the yellow caudal keels, a diagnostic feature of this species.
marinethemes.com/ Lawrence Martin

LONGTAIL TUNA
Thunnus tonggol

OTHER COMMON NAMES

	Oriental bonito
	Tonggol tuna
Australia	Northern bluefin tuna
France	Thon mignon
Japan	Koshinaga
Spain	Atún tongol

The longtail tuna is a very important food fish in some parts of the world, and is also widely regarded in sport fishing circles as perhaps the toughest of all the tunas on matched angling gear. Unfortunately, it is one of the least studied of all of the tunas.

The longtail tuna is the only tuna which lacks a swim bladder. This feature requires dissection and some anatomical knowledge, so perhaps an easier internal feature to note is the presence or absence of small dark striations on the surface of the liver. The longtail lacks these, while southern bluefin and bigeye tuna both have these characteristic striations. On the other hand, the yellowfin tuna also has a plain liver, but possesses a relatively long pectoral fin, usually reaching to at least the middle of the base of the second dorsal, whereas the pectoral fin of the longtail is relatively short, rarely reaching to the space between the first and second dorsal fins. Lastly, longtail tuna are not nearly as brightly colored as yellowfin, having only a tinge of yellow in the two dorsal fins. The longtail also lacks the 'dot-dash' belly stripes of the yellowfin, instead possessing only a few scattered pale spots on the belly and lower flanks (only evident on freshly caught fish).

Whereas most of the true tuna species have very broad, often transoceanic distributions, the longtail tuna's habitat and range is largely restricted to the coastal waters of the northern half of Australia, most of Southeast Asia and the coastal regions of the northern Indian Ocean. Even though the longtail is a coastal species, it tends to avoid turbid or 'dirty' water, and rarely enters estuaries, although it is often found in large marine-dominated embayments.

The only place where longtail tuna have been tagged in numbers is eastern Australia. Just over 4200 longtail have been tagged and released there, and 57 recaptures have been reported. All movements so far revealed have been along the coast, some at very rapid rates (for example, 550 km in 34 days and 850 km in only 20 days). A genetic study of the stock structure of the species throughout its range, together with some satellite tagging operations, would be important research priorities for this species.

The growth rate of longtail tuna is not well understood. Several studies using different methods have had varying results. The best estimates of age at size are generally agreed to be about 45 cm for one-year-old fish, 73 cm by age two, 92 cm by age three and 105 cm by four years of age. Recent studies suggest the species may be quite long-lived, with estimates for some specimens as old as 19 years.

Significantly, all of the line class IGFA world records for longtail tuna have been taken at the southernmost edge of the

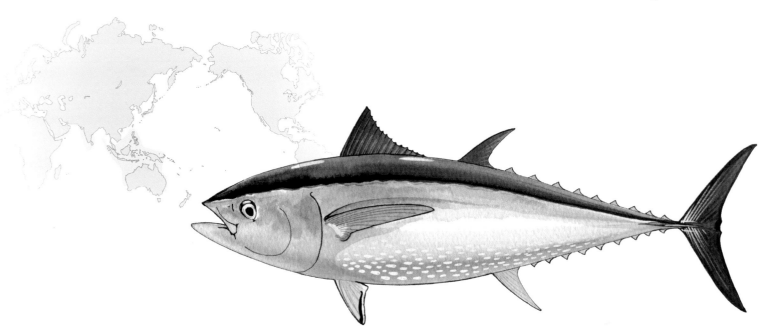

species' range, off southeastern Australia. The maximum recorded size is a 35.9 kg (79 lb) specimen caught at Montague Island, Australia, in 1982. Fish over 30 kg (66 lb) are uncommon and seem to be rare in commercial catches anywhere else in the world, so the population of large longtails found seasonally on the Australian southeast coast may well be very important to the overall stock since these are apparently prime, adult fish.

Studies in the Indian Ocean have indicated that longtail tuna mature at a very small size; in fact, as small as 45 to 50 cm, or an age of only one year. On the other hand, research conducted in Papua New Guinea found size at first maturity to be between 60 and 70 cm (about two years old according to the growth studies mentioned above). While it is thought by some scientists that spawning may occur twice per year, it is now becoming increasingly apparent that all of the tunas probably spawn on almost a daily basis during extended spawning seasons. It is therefore highly likely that the longtail does the same, although this is yet to be proven.

The longtail tuna does not usually form large schools like other tropical tunas. Small schools of longtail tend to surface only for brief periods, and are generally very 'flighty', being easily scared into sounding or scattering. An interesting observation is that adult longtail tuna, in pods of 15 to 20 equally spaced individuals, often attack their prey in a parabolic-shaped formation. This type of cooperative feeding behavior has also been observed in giant bluefin tuna in the Atlantic.

Because the longtail is primarily a coastal species, it tends to be the target of many small-scale coastal fisheries throughout its extensive range through many countries. The total catch in the Indian Ocean in the 1980s was only about 80 000 tonnes, but this has increased to more than 250 000 tonnes in the last few years. The bulk of this catch consists of small fish, mostly between 50 cm and 80 cm in length – about 2 to 8 kg (5 to 17 lb) – and while some of these fish would have reached spawning age, many would not. Most other tuna species are oceanic in their distributions, which means that they lend themselves to reasonable international management at a high level. Unfortunately, this is not the case for longtail, which remains the forgotten tuna in international fisheries forums.

BLACKFIN TUNA
Thunnus atlanticus

OTHER COMMON NAMES

France	Thon à nageoires noires
Japan	Mini maguro
Spain	Atún aleta negra

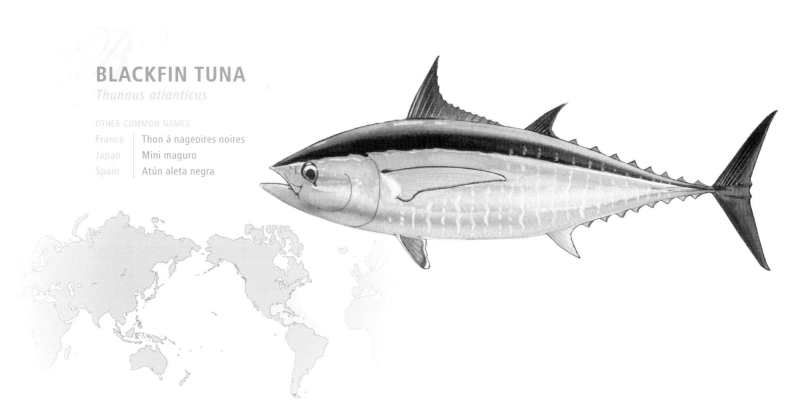

Of the eight species of true tuna (that is, those fishes belonging to the genus *Thunnus*), the blackfin tuna is probably the least studied. Confined to the western Atlantic, and of minor importance commercially or recreationally, this relatively obscure tuna poses some fascinating questions regarding its evolution and zoogeography. It is closely related to, and overlaps with the yellowfin tuna, yet maintains its distinct species status, albeit, within a remarkably limited distribution.

The blackfin tuna is similar to the yellowfin tuna in general appearance, and in the possession of a plain colored liver. However, it can be generally distinguished by the color of its fins. Whereas the second dorsal fin and the finlets of the yellowfin tuna are yellow, the second dorsal of the blackfin is dark silvery gray while its finlets are also dark, with just a touch of yellow. As a final check, blackfin tuna have 19 to 25 gill rakers on the first gill arch, while yellowfin have 26 to 34.

This tuna species has the most limited distribution of any of the genus. It is entirely confined to the western Atlantic Ocean, ranging from the northern coast of the United States to about the latitude of Rio de Janeiro, Brazil, and encompassing all of the Caribbean islands. Blackfin tuna prefer temperatures above 20°C.

Virtually nothing is known about the movements of blackfin tuna. The highly restricted distribution of the species strongly suggests that individuals do not undertake lengthy movements or migrations, although it is thought that spawning grounds are well offshore. Being an oceanic, epipelagic fish, it is somewhat of a mystery why it has not spread beyond its current range, given that all of the other *Thunnus* species have very broad distributions.

This is the smallest of the *Thunnus* species. The maximum size to which blackfin tuna grow is about 100 cm, or around 20 kg (44 lb) – the all-tackle

world record is 22.4 kg (49 lb). However, they are mostly caught at sizes of 60 to 70 cm, or 5 to 7 kg (11 to 15 lb) in weight. One of the few studies on growth rates of blackfin tuna (conducted at Martinique) estimated that they attain a length of about 32 cm by seven months, and 70 cm by three years of age.

Little work has been carried out on the reproduction of blackfin tuna. Spawning fish are found off Florida from April to November, while inside the Gulf of Mexico spawning is thought to occur from June to September only.

Compared with other tunas, the blackfin is not particularly important commercially. It is caught by pole-and-line, and by various artisanal methods throughout the Caribbean. Often forming mixed schools with skipjack tuna, blackfin may be regarded as an incidental species which is not separated in skipjack tuna statistics of the western Atlantic.

SKIPJACK TUNA
Katsuwonus pelamis

OTHER COMMON NAMES

	Striped tuna
	Oceanic bonito
France	Bonite à ventre rayé
	Listao
Japan	Katsuo
Spain	Listado

Known by many names around the globe, the skipjack is one of the world's most important food fishes. It is by far the dominant species caught in the western and central Pacific Ocean tuna fisheries and is the main species used in tuna canneries around the world. The skipjack is also a very important recreational species, providing sport, food and prime bait for most of the larger predatory fishes.

The skipjack is easily recognized by its horizontal belly stripes. It is sometimes confused with the bonitos, but the latter have horizontal stripes over most of the body, and also possess sharp conical teeth, which the skipjack lacks. In life, the dark back of the skipjack ripples with seemingly fluorescent, purple hues, while the white undersurface reflects pinks and mauves. Another transient color variation which skipjack often exhibit is to produce vertical bars when excited or feeding. These bars break up the permanent horizontal stripes, perhaps signaling to others in the school, generating a feeding frenzy and resulting in confusion and panic in their target prey species.

The skipjack tuna is one of the most widely distributed of all pelagic fishes. Its range extends at least between the latitudes of 40°N and 40°S in all three major oceans. It is found to the south of the Australian and African continents, and it is therefore widely assumed that the species mixes between major oceans, although extensive tagging is yet to prove this speculative point. The widespread distribution is primarily due to the skipjack's broad thermal tolerance, from less than 15°C to over 30°C. It has long been known that skipjack concentrate at convergences of water masses, and in recent years, sophisticated computer models have shown that there is a very strong correlation between skipjack abundance and planktonic concentrations (as indicators of productive waters). In addition, the models have demonstrated marked shifts in the distribution of skipjack across the tropical Pacific depending on the El Niño/La Niña cycle. In El Niño years, the center of concentration of skipjack in equatorial waters shifts eastwards, while in La Niña years, the shift is back towards the west, particularly around eastern and northern Papua New Guinea where catches are among the world's highest.

For a species which is so incredibly important to the economies of many small countries, it is surprising that the growth rates of skipjack are not known with greater certainty. This is partly because the growth rate almost certainly varies greatly from place to place, depending on food availability, but nevertheless,

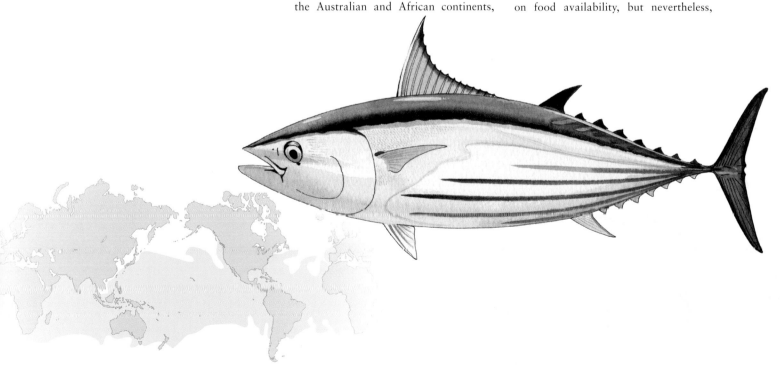

few studies have been conducted. The best estimates, based on tag recapture data and examination of growth bands on bony parts, are that skipjack attain a spawning size of about 45 cm by their second year of age and probably only live for a maximum of four years.

The maximum size of skipjack quoted in the scientific literature is 108 cm fork length, which would correspond to a weight of about 33 kg (73 lb). No fish approaching this size has been caught recreationally, the IGFA all-tackle record for skipjack being 20.5 kg (45 lb) for a fish caught off Baja California in 1996. One nearly as large – 19 kg (42 lb) – was caught off Mauritius in the Indian Ocean in 1985, while another Mauritian specimen caught in 2004 weighed 18.9 kg (41.6 lb). The largest Atlantic skipjack listed is a 17.8 kg (39 lb) fish caught off Bermuda in 1978.

Like many other aspects of its biol-ogy, the reproduction of the skipjack is quite extraordinary. This was the first tuna species proven not only to produce huge numbers of eggs, but to spawn every day (or thereabouts) for months on end. The number of eggs spawned each day by a single female skipjack ranges between about 100 000 and 2 000 000, which, considering the sheer biomass of skipjack tuna in the tropical oceans, is a mind-boggling concept. Is it any wonder that this little tuna has sometimes been dubbed 'the cockroach of the sea'?

Greater numbers of skipjack tuna have been tagged than any other pelagic species. In expensive, dedicated tagging cruises lasting months at a time, hundreds of thousands of skipjack have been tagged throughout large parts of their range in the western Pacific. Nearly all of these have been captured by the pole-and-line method, whereby a school of fish is located, and kept feeding at the surface by chumming, preferably with live bait. Fish are then hooked by a simple pole-and-line armed with a barbless jig, hoisted over the shoulder, whereupon a team of technicians carefully measure the fish in a soft cradle, tag it and release it, all within 30 seconds.

These tagging operations have primarily been designed to assess the stock status of skipjack; that is, the exploitation rate compared with the standing biomass. However, the other spin-off of the tagging has been the study of rates and extent of movements of the species through its range. Through these studies, skipjack have been shown to be highly mobile fishes, tending to travel further and further the longer they are at liberty. Movements are apparently relatively random, with gradual mixing of stocks over very large areas throughout the species' range.

The fisheries for skipjack tuna in the world's oceans are huge. Two main fishing methods are used: purse seine and pole-and-line, with the former taking the bulk of the catch. Many of the purse seine vessels which target skipjack in all oceans are rightly called 'super-seiners'. These huge boats, often equipped with their own helicopters, as well as a vast array

of electronic fish-finding gear, travel great distances in search of the prolific skipjack. When a school is found, jet boats are launched immediately to deploy the million-dollar nets as quickly as possible. A single net can encompass a school literally the size of a football ground, and more. At the bottom of the net is a rope which is then pulled until the net is closed, or 'pursed', after which the net is gradually brought to the boat and the catch scooped out. Purse seiners mainly set their gear in two ways: either on free-swimming schools of tuna ('non-associated sets') or around floating objects ('associated sets'). Sets on free-swimming schools tend to take almost exclusively skipjack, yellowfin and bigeye tuna, but sets on floating logs, and increasingly on fish aggregating devices (FADs) deployed by the vessels, take a large amount of bycatch. The bycatch in these cases is not commonly marine mammals, but the whole fish community which is associated with the floating object, including, among others, billfish and sharks. This is a trend which is of considerable concern, and which will need to be closely monitored.

The quantity of skipjack tuna taken in the western and central Pacific has increased steadily, from about 400 000 tonnes in 1972 to over 1.4 million tonnes in 2002 to 2.4 million tonnes in 2007. The latest figure represents 84 per cent of the total Pacific Ocean catch and 55 per cent of the global tuna catch (the provisional global estimate for 2007 is just under 4.4 million tonnes). Increases in technology, including deployment of thousands of fixed or expendable fish aggregating devices fitted with satellite beacons, and the use of bird radar, which can detect flocks of seabirds over 20 miles or more, have contributed to the increased world catch. But the question which is continuously asked is: what is the ultimate limit to the sustainable catch of skipjack? As noted in the section on reproduction, this species is 'built' for proliferation and fishery models indicate that the stocks are sustainable at the current levels of exploitation and are not overfished.

A secondary question regarding this huge take of biomass out of the ocean relates to the possible effects this may (must?) be having on the marine food chain both in terms of skipjack prey species and, at the other end, on the predators of skipjack such as the billfishes. Most aspects of these questions have rarely been seriously studied scientifically, and predicting effects of food availability or oceanographic variables on absolute abundance (rather than effects on aggregation, or local abundance) is a major hurdle. Computer models are being developed to attempt to grapple with such apparent imponderables, but considering all the natural complexities, they still have a considerable way to go to accurately simulate the real world and, thereby, answer such questions.

LITTLE TUNAS

Euthynnus affinis,
E. alletteratus, E. lineatus

	Mackerel tuna
	Kawakawa
	Bonito
	Black skipjack
	Little tunny
France	Thonine
Japan	Hiragatsuo
	Yaito
Spain	Bacoreta
	Barrilete

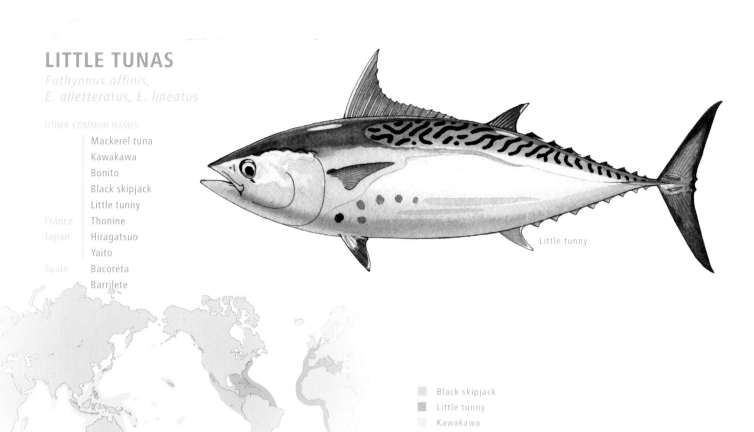

Little tunny

Black skipjack
Little tunny
Kawakawa

The three species of so-called little tunas occupy similar niches in their respective broad regions around the world. The Indo-western Pacific species, *Euthynnus affinis*, is known as the kawakawa or mackerel tuna, while the closely related species in the eastern Pacific Ocean, *Euthynnus lineatus*, is known somewhat misleadingly as the black skipjack. (This is misleading since it is not a true skipjack tuna, of which there is a single worldwide species, *Katsuwonus pelamis*. Therefore, perhaps a better name for this fish would be the black kawakawa.) The fact that the little tunas have evolved into two distinct species on either side of the Pacific strongly indicates that these species are not nearly as migratory as the true skipjack, which is genetically uniform throughout its huge range. In the Atlantic, the little tunas are represented by a single species, *Euthynnus alletteratus*, known as the little tuna, or little tunny.

Wherever they occur, the little tunas are often the first tuna species to be encountered by novice sportfish anglers. This is because the species comes into range of small boat anglers, tending to be coastal and even entering large embayments and estuaries from time to time.

All three little tunas are distinguished from true tunas (*Thunnus* spp.) and from the skipjack tuna by the fact that their bodies are scaleless (naked) behind the corselet – a heavily scaled area surrounding the head and 'shoulders'. In comparison, the area behind the corselet is covered in small deciduous scales in the skipjack and true tunas.

The little tunas are characterized by broken wavy horizontal markings confined to the dorsal area behind the corselet. All three species also nearly always have several distinct dark or black spots on the breast area, between the bases of the pelvic and pectoral fins. This is a useful feature to separate them from the similar-looking bullet or frigate tunas (*Auxis* spp.). Another feature which separates the bullet and little tunas is the pointed process situated between the paired pelvic fins (the ones under the breast). This process in bullet tunas is single, having just one point, but is two-pointed in the little tunas.

The three species of little tunas are largely separate in their distributions. However, the two Pacific species, kawakawa and black skipjack, may very

occasionally be found in the same areas. They can be distinguished primarily by gill-raker counts. The kawakawa has 29 to 33 gill rakers on the first gill arch, while the black skipjack has 33 to 39.

The kawakawa (*E. affinis*) is largely confined to continental shelves and islands of the western Pacific and Indian oceans. It is an unusual distribution in that there is some continuity between widely separated archipelagos, indicating that the species does occur in the open ocean, at least to some extent. This continuity breaks up in the central Pacific, where the kawakawa is found only in association with major island groups such as the Hawaiian chain and French Polynesia. The occasional specimen has turned up off the Californian coast, but so rare is that event, that it has generated special attention in the scientific literature.

The black skipjack (*E. lineatus*) only occurs in the eastern tropical Pacific, from southern California to Peru, including the Galapagos Islands. Its distribution also extends occasionally to Hawaii, the only area where its range might overlap with that of the kawakawa with any regularity.

The little tunny, *E. alletteratus*, is

confined to the tropical or subtropical continental margins on either side of the Atlantic Ocean, also extending into the Mediterranean Sea.

A surprising number of kawakawa have been incidentally tagged under the Australian Gamefish Tagging Program. More than 18000 have been tagged off the east coast of Australia, and 60 recaptures have been reported. All recorded movements have been relatively short (less than 500 km) and coastal, with no fish showing evidence of moving out into, or across oceanic waters. There is no information available on movements of either the black skipjack or little tunny, although because of their relatively restricted distributions, it tends to be assumed that they are not migratory in any real sense, even though they are classified as 'highly migratory' under international conventions.

The IGFA all-tackle record for kawakawa is 13.15 kg (29 lb), although there must be some doubt about this particular fish since it was caught in 1986 in the Revillagigedo Islands, off the Mexican coast in the eastern Pacific. As noted, the kawakawa is very rare in the eastern Pacific, so it is likely that this fish was a black skipjack (although this can't be proven). If that is the case, then the largest certifiable kawakawa on record would be one which weighed 11.04 kg (24.3 lb), caught off South Africa in 1995. That fish just bettered an 11.00 kg (24 lb) fish caught off eastern Australia in 1987. The official record for black skipjack is 11.8 kg (26 lb), for a fish caught off Baja California, Mexico, in 1991. The little tunny grows to a somewhat larger size than its two congeners. The largest of that species was a 16.32 kg (36 lb) specimen caught off New Jersey in 2006.

Spawning seasons of little tunas vary from place to place, usually covering two to three months of the year. A female kawakawa weighing only 1.4 kg (3 lb) may spawn over 200000 eggs per batch (presumably like the skipjack, on a daily basis), while a 5 kg (11 lb) fish would spawn nearly a million eggs at a time. Like other tuna species which are frequent spawners, the largest little tunas are males since much of the energy of the females is assigned to constant egg production rather than growth.

There are no large-scale fisheries for the little tunas, although they often school with other species of tuna at similar sizes, including skipjack (*Katsuwonus pelamis*), and are therefore caught in association with the skipjack fishery. However, because they constitute only a small proportion of the total catch, they are not recorded separately, so the true size of the catch is unknown.

Because of their association with land masses, the little tunas can be important components of coastal artisanal fisheries. For example, it is quite common to see kawakawa at fish markets through its range in Southeast Asia, where it is mainly caught as a bycatch of coastal gillnet fisheries that primarily target narrow-barred Spanish mackerel (*Scomberomorus commerson*).

BULLET TUNA

Auxis rochei, A. thazard

Bullet tunas are the smallest members of the true tuna tribe, Thunnini. They swim under an amazing variety of names, depending on where you find yourself at the time and who you are talking to. They are variously called bullet tuna, bullet mackerel, frigate tuna, frigate mackerel, horse mackerel and, oddly, leadenall. The last name is Australian and most likely refers to the metallic lead-like color on the belly and flanks of live fish.

There are two worldwide species of bullet tunas, the bullet tuna, *Auxis rochei* and the frigate tuna, *Auxis thazard*. As well, eastern Pacific subspecies of each have been described. However, for simplicity, we will consider the two main forms here.

Bullet tunas are most likely to be confused with small little tunas (kawakawa, little tunny and black skipjack) since both groups have permanent stripes or wavy markings on their rear dorsal surfaces. However, frigate tuna nearly always lack any spots along their forward flanks, whereas all three species of little tuna always have several such spots on either side. Another good identifying feature of the bullet tunas is the possession of a single central, pointed process in between the paired pelvic fins (the ones under the breast). In the little tunas the process has two distinct, backwardly directed points.

The bullet tuna (*Auxis rochei*) is generally more slender than the frigate tuna (*Auxis thazard*), which has more of the robust shape of a miniature tuna. The two can also be distinguished by the width and length of the corselet of scales and body pattern. In the bullet tuna, the rear part of the corselet is wide, while it is narrower in the frigate tuna. Also, the patterning on the back of the bullet tuna shows almost vertical barring compared with more oblique to horizontal wavy stripes on the frigate tuna.

Both of the bullet tuna species have a worldwide distribution, and overlap considerably throughout their ranges (as do the two eastern Pacific subspecies). They tend to be largely confined to continental shelves and to some island groups in each major ocean and to have broader temperature tolerance than other small tuna, with the frigate tuna, for example, extending, around the southern coast of

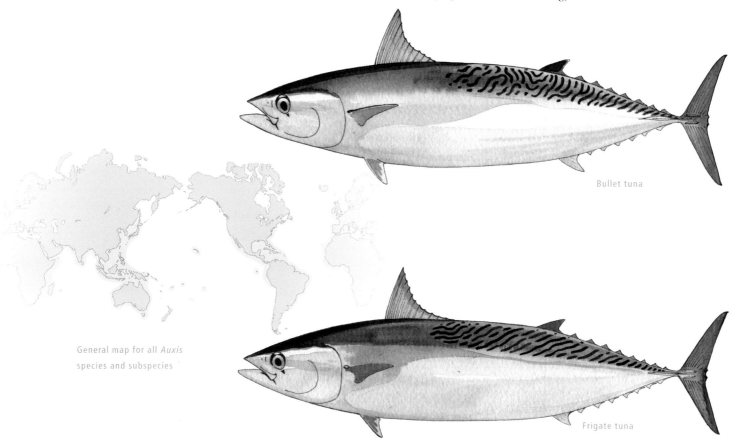

General map for all *Auxis* species and subspecies

Bullet tuna

Frigate tuna

Australia and the South Island of New Zealand.

The maximum size of the frigate tuna referred to in the scientific literature is 58 cm, which would be perhaps 4 kg (9 lb) in weight. This would be an exceptional size, though, and average fish would tend to weigh between 1 and 1.5 kg (2 to 3 lb). Because of its small maximum size, the species is not recognized as a gamefish by IGFA, so no angling records are available.

Bullet tunas tend to be spring or fall spawners. As is the case for all of the tunas, frigate tunas are highly fecund. A female fish can produce between 30 000 and 100 000 eggs per spawning according to the size of the fish, and spawning probably occurs every day or every other day during the spawning season. Body length at first maturity is about 35 cm. Very few bullet tuna have been tagged and we are therefore ignorant about their movements. However, given the species' streamlined body form, their geographical range and seasonality, it would not be surprising to find that these are highly mobile fish capable of long-distance movements.

Bullet tunas are enigmatic fishes which are probably caught in quite large quantities but because they are often a part of a mixed catch, are not recorded as separate species. Even so, the reported world catch of *Auxis* species is more than 200 000 tonnes per year. Also, like the little tunas, bullet tunas are important food fish for coastal communities in developing countries where catch statistics are largely unrecorded.

The global biomass of bullet tunas is likely very large and as such, these are very important components of the marine food web, especially as prey for apex predators such as billfishes and large tunas. I have personally inspected the contents of the stomachs of hundreds of marlin, and one species which is a consistent dietary item, at least of black marlin caught off southeastern Australia, is the frigate tuna.

BONITOS

Sarda spp.,
Cybiosarda elegans,
Orcynopsis unicolor,
Allothunus fallai

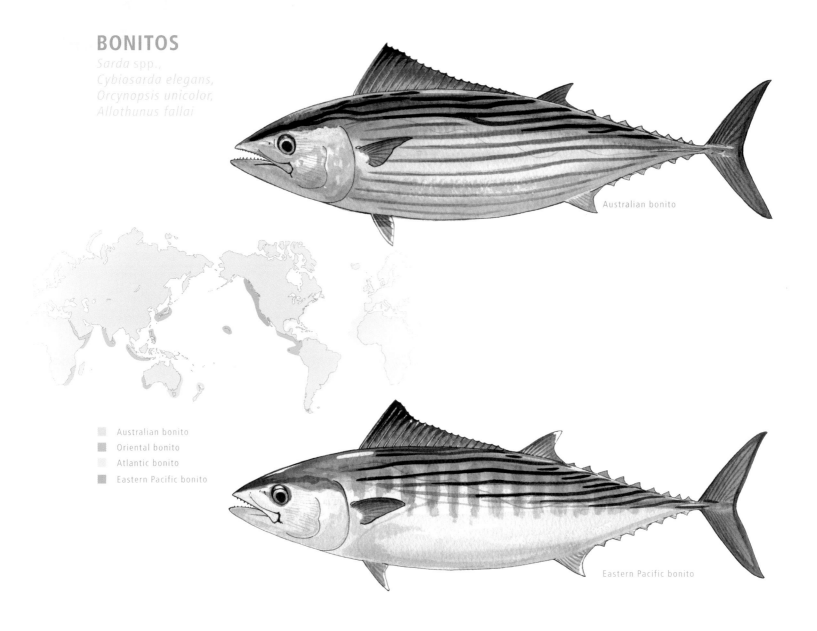

Australian bonito

Australian bonito
Oriental bonito
Atlantic bonito
Eastern Pacific bonito

Eastern Pacific bonito

The bonitos are members of the family Scombridae, belonging in turn to the Sardini – the tribe most closely related to the tuna tribe, Thunnini. The main feature separating the bonitos from the tunas is a technical one – bonitos lack cartilaginous ridges on the tongue. The bonitos do have prominent teeth, however, which are more or less conical in shape, rather than compressed and knife-like, as in the mackerels (*Scombero-morus* spp.). One member of the bonito tribe, the dogtooth tuna (*Gymnosarda unicolor*) is treated separately next. The others are briefly described below. All are small scombrids – less than 15 kg (33 lb) – with mainly coastal distributions, although several have broader ranges. The term 'striped' bonitos, as used below, refers to the four *Sarda* species.

Atlantic bonito
Sarda sarda

Occurring as it does along the coasts of northern Europe, including the Mediterranean and the Black Sea, this bonito species has been known and fished since antiquity. It is also common along the northeastern United States, extending to the Gulf of Mexico. Its total distribution is broad and includes the entire west coast of the African continent, and a population off Venezuela and Columbia.

Of the four striped bonitos, the Atlantic bonito has by far the most oblique stripes along its back and flanks, distinguishing it from all other possibly similar looking species.

This is a relatively small bonito, apparently not reaching the sizes attained by the other striped bonitos. The IGFA all-tackle record fish weighed 8.3 kg (18 lb) and was caught in the Azores in 1953. Another weighing just over 8 kg (17.6 lb) was taken off the Canary Islands in 1993.

Australian bonito
Sarda australis

Also known as the horse mackerel, the Australian bonito is readily recognized by its horizontal stripes which cover the back, and usually extend down the flanks where they tend to break up into dotted lines. This species is sometimes confused with the skipjack tuna, but the latter only has stripes on its lower half, and lacks the sharp, conical teeth of the true bonito.

The Australian bonito is endemic to that part of the world, being restricted entirely to the southeast coast of the continent, including Tasmania, with some scattered records also noted from Norfolk Island and the North Island of New Zealand.

Spawning is unknown, but sexual maturity is thought to be reached in this and the other striped bonitos by a size of about 40 cm. The growth rate of the Australian bonito is completely unstudied, but it is likely that this is a fairly short-lived fish, if only because recaptures of tagged fish after more than one year are quite rare.

The IGFA all-tackle record for the species is a 9.4 kg (20.7 lb) specimen taken in 1978.

Oriental bonito
Sarda orientalis

The oriental bonito has the broadest and perhaps most unusual distribution of the striped bonitos. It hugs the northern rim of the Indian Ocean, but also occurs in the Pacific, off Japan and the Philippines, in the Papuan Gulf, around the Hawaiian islands, and along the Californian, Central American and Peruvian coasts, and as far south as the Galapagos Islands.

The oriental bonito is readily distinguished from the Australian bonito by its gill-raker count (the hard, comb-like processes on the front of each gill arch). The oriental bonito has between eight and 13 of these gill rakers, while the Australian bonito has 19 to 21. Also, the oriental only has horizontal stripes on the top part of the body, whereas the Australian bonito has them over most of the lateral surface.

Oriental bonito school with other species of small tuna, especially kawakawa and longtail tuna, and are caught in artisanal fisheries throughout their range by trap, pole-and-line, gill net and trolling. Like many of the tunas, spawning takes place in summer months.

The maximum reported size in the scientific literature for oriental bonito measured a surprising 101.6 cm, but no weight was given for that specimen. The IGFA all-tackle record was a 10.65 kg (23 lb) fish measuring 89.5 cm fork length caught in the Seychelles in 1975, so a 100 cm plus specimen would be considerably heavier.

Eastern Pacific bonito
Sarda chiliensis chiliensis,
S. chiliensis lineolata

Until recently, only one eastern Pacific bonito was recognized – *Sarda chiliensis*. The species has since been split into two subspecies, a southern form (*Sarda chiliensis chiliensis*) found from Central America to Chile, and a northern variety (*Sarda chiliensis lineolata*) which occurs, rather surprisingly, from Alaska, through coastal southern California to Baja Mexico. Both are characterized by slightly oblique horizontal stripes confined to the dorsal surface.

The maximum sizes for both forms are given in the literature as 102 cm in length and 11.3 kg (25 lb) in weight, although this is cited as a northern hemisphere fish. The IGFA all-tackle record was also the northern form, caught off California in 2003 and weighing 9.67 kg (21.3 lb).

Leaping bonito
Cybiosarda elegans

This beautiful little member of the bonito tribe, also known as Watson's leaping bonito, is restricted to the Australian east coast, north of about 35°S, extending through Torres Strait to the southern coast of Papua New Guinea. It is a distinctive species which can be distinguished from the other bonitos in having both horizontal striping on the flanks, and black elongated spots on a dark blue-green background on the upper half of the body. It is also characterized by a very high first dorsal fin which is mostly jet black, but with a white patch at the rear. When feeding in shoals, as their name implies, they leap from the water, coming from below to attack their small prey. This is the smallest of the bonitos, only growing to a size of 2 to 3 kg (4 to 6.5 lb).

Plain bonito
Orcynopsis unicolor

The plain bonito is well named. It is a small scombrid with few distinguishing body markings, other than a scattering of faint spots on the dorsal surface and a dorsal fin which is black on its front half. Its range is quite restricted, occurring only along the coast of western Africa, north of the Congo, and inside the Mediterranean. It is not a common species and very little is known about its biology. It grows to a maximum size of about 13.1 kg (29 lb), and a cited length of 130 cm.

Slender tuna
Allothunnus fallai

The slender tuna is misnamed in that it is really a member of the bonito tribe. This species stands alone from the rest of the group, however, in that it is by far the most oceanic and the least tropical in its distribution. It is a pelagic species restricted to a narrow circumglobal band between the latitudes of about 20°S and 50°S. It is not common throughout this range, but on occasions, catches of up to 80 tonnes have been made by purse seine.

This pretty fish has the typical pelagic coloration of many oceanic fishes, with a dark blue back and pale to white flanks and belly. It has no distinctive markings such as bars or spots.

The slender tuna grows to a little under 1 metre in length, and a maximum weight a little under 10 kg (22 lb).

A rare photo of a large school of Watson's leaping bonito. This is the smallest of the bonitos, and perhaps the prettiest.
marinethemes.com/ Erika Antoniazzo

DOGTOOTH TUNA
Gymnosarda unicolor

OTHER COMMON NAMES

	Scaleless tuna
	Lizard-mouth tuna
France	Thon blanc
Japan	Isomaguro
Spain	Bonito sierra

Within the family Scombridae, the closest relatives to the tunas (tribe Thunnini) are the bonitos (tribe Sardini). As noted, this tribe contains four closely related species of striped bonitos (genus *Sarda*), and three other species – the 'plain' bonito of the Mediterranean, Watson's leaping bonito, and the dogtooth tuna. This means that, technically, the dogtooth tuna is not really a tuna at all, but rather, a bonito – and a giant bonito at that. The principal features that make the dogtooth a bonito are the lack of hard, longitudinal ridges on the tongue (which tunas possess), and prominent conical teeth.

The dogtooth tuna is a distinctive species that could never be confused with any other. It is plain green to gray on the dorsal surface with no stripes or spots on the body, has about 20 large, sharp, conical teeth in both its upper and lower jaw, and bears a single, prominent wavy lateral line.

The distribution of the dogtooth tuna is particularly interesting, reflecting its special habitat preference for reef drop-offs. The outer reefs of northern Australia appear to be the center of its distribution, where it occurs through the extensive reef system of the Coral Sea, extending through to eastern Papua New Guinea and the Solomon Islands and forming isolated populations around remote coral atolls. Its patchy distribution in the Pacific includes the Philippines, Japan and French Polynesia (Tahiti and surrounding islands) as well as Fiji, Vanuatu and Pitcairn Island. In the Indian Ocean, it is found on the northwest shelf of Australia, in some parts of Indonesia, off India and Sri Lanka, around Madagascar and, somewhat surprisingly, in the Red Sea. The dogtooth is a tropical species which prefers water over 25°C, but is sometimes found in cooler climes.

Virtually nothing is known about the movements of dogtooth tuna. The dogtooth tuna is sometimes called a migratory species, but considering its restricted distribution, and knowing the habitat preferences of adults, it could be fairly safely assumed that most of the dispersal occurs at the larval, postlarval or juvenile stages. Adult dogtooth continually patrol the vertical reef drop-offs and divers consider them to be territorial in some areas.

Virtually no work has been undertaken on growth rates of dogtooth. One large specimen of 75 kg (165 lb) was recently aged via its otoliths (ear bones)

and estimated to be 19 years old. If so, this would indicate a particularly slow growth rate for a scombrid.

The all-tackle record for dogtooth tuna weighed 104.5 kg (230.5 lb) and was caught near the Indian Ocean island of Mauritius in 2007, while the next largest specimen in the record charts was also caught at Mauritius (in 1993) and weighed 104.3 kg (230 lb).

It is a little surprising that dogtooth tuna larvae have rarely been collected in surveys, although most of these surveys have been undertaken near inshore reefs. It is therefore likely that spawning mainly takes place around the more remote coral atolls, the locations where the larger adults are normally found.

The little work that has been done on reproduction of dogtooth tuna suggests that the size at first maturity is about 65 cm fork length, which is quite small for a species which grows to at least 150 cm.

There are no specific commercial fisheries which target dogtooth tuna, although they are targeted and caught by Pacific Islanders for food. In some areas, large specimens have been implicated with ciguatera poisoning, but this is not common and dogtooth are regarded as safe to eat. The main 'commercial' fishery for this unique fish would therefore be the adventure fishing industry which promotes dogtooth fishing trips to locations such as Papua New Guinea, Fiji and Vanuatu, as well as long-distance trips from Australia to remote atolls in the Coral Sea.

There is some suggestion that dogtooth tuna can be fished down in an area fairly quickly, especially with respect to taking out the larger specimens. Much of the evidence of declines in numbers of dogtooth tuna is anecdotal, but nevertheless, certainly warrants a specific research program to investigate the life cycle of this unusual giant bonito and effects of fishing on its stocks.

BLUE MACKERELS
Scomber spp.

OTHER COMMON NAMES

	Slimy mackerel
	Chub mackerel
	Striped mackerel
France	Maquereau
Japan	Saba
Spain	Caballa

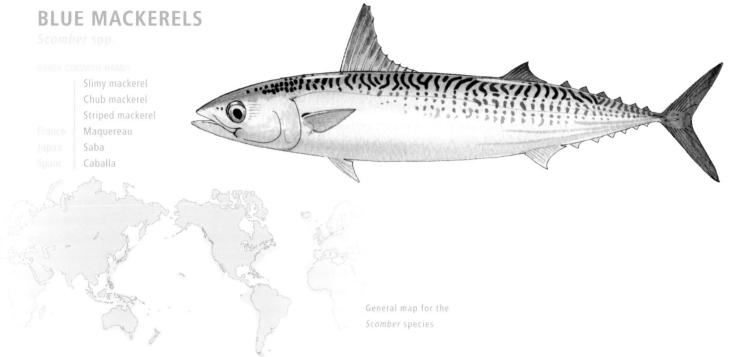

General map for the
Scomber species

Blue mackerel are, in essence, miniature tuna. They are members of the family Scombridae, which also includes all of the tunas and larger, toothed mackerels. They are not, however, close relatives of the true 'Spanish' mackerels, but belong to a tribe called the Scombrini, which also includes the Indian mackerels (*Rastrelliger* species). Blue mackerel can be distinguished from the Spanish mackerels by their lack of sharp teeth and their numerous, filamentous gill rakers which are used to strain plankton from the water – their principal mode of feeding. They can be separated from their closer relatives, the Indian mackerels, by their round bodies (Indian mackerels are strongly laterally compressed).

Worldwide, there are four species of blue mackerel, all in the same genus, *Scomber*. These are the blue or slimy mackerel, *Scomber australasicus*, the chub mackerel, *Scomber japonicus*, the Atlantic chub mackerel, *Scomber colias*, and the Atlantic mackerel, *Scomber scombrus*. All three are very similar in body form and patterning, but have markedly different geographic distributions.

All of the blue mackerels range widely in shelf-associated temperate and subtropical oceans and seas. They occur to the edge of continental shelves, but are rare beyond.

They are small fishes, often lumped under the general term 'baitfish', which alludes to their importance as forage fish for larger predatory fishes and other marine animals. Since their diet is mainly planktonic, and because their biomass is so large, the blue mackerels are extremely important links in the marine food web, between plankton and predators such as tunas, marlin, dolphins and seals.

Blue mackerels are batch spawners, producing up to 450 000 eggs per female each season. They are surprisingly long-lived for such small fishes, attaining maximum ages of at least ten years and up to 18 years old. Inshore fish tend to be juveniles, usually smaller than 30 cm in length. Adults are found in deeper, offshore waters where maximum sizes attained range from about 50 cm and 1.3 kg (3 lb) for Atlantic mackerel to 65 cm and 2.9 kg (6.4 lb) for chub mackerel.

Blue mackerel form the basis of substantial fisheries around the globe, with harvests totaling at least several million tonnes per annum. The flow-on effects of such catches on populations of higher level predators is unknown at present.

INDIAN MACKERELS
Rastrelliger kanagurta,
R. brachysoma, R. faughni

OTHER COMMON NAMES
    Short mackerel *R. brachysoma*
    Island mackerel *R. faughni*
    Longjaw mackerel *R. kanagurta*
France   Maquereau des Indes
Japan   Gurukuma
Spain   Caballa de la India

The prolific blue mackerel is a vital component of the open-oceanic food web.
Phillip Colla/
oceanlight.com

The three species of Indian mackerels are members of the Scombrini tribe of tunas and mackerels. They possess long, filamentous gill rakers with fine side bristles, which are used to filter phytoplankton, a feature they share with the blue mackerels, *Scomber* spp. They are true filter feeders with large gaping mouths which they hold open while feeding in slow-moving schools.

The Indian mackerel (*Rastrelliger kanagurta*) is the most widespread of the three, occurring throughout the Indo-west Pacific, from about the Red Sea in the west of its range to Samoa in the east. It has been reported as entering the Mediterranean Sea via the Suez Canal. The short mackerel (*R. brachysoma*) and the island mackerel (*R. faughni*) both occur throughout Southeast Asia, extending around Papua New Guinea and Fiji. All are pelagic species which are largely neritic; that is, they occur in relatively shallow, coastal waters.

With the three species overlapping in their distributions, a question arises as to how each maintains its integrity. One answer lies in their diets. The short mackerel has 30 to 48 gill rakers on the lower part of the first gill arch. It feeds on micro zooplankton and phytoplankton; that is, the smallest plankton of the group. The Indian mackerel has about the same number of gill rakers (30 to 46) and filters medium-sized plankton – primarily larval crustaceans. The island mackerel has the least number of gill rakers (21 to 26) and feeds on the largest plankton, large zooplankton.

These are small mackerels growing to a length of about 20 cm (island mackerel) to 35 cm (Indian and short mackerels). They are most likely relatively short-lived fishes; Indian mackerel, for example, being thought to live for only four years.

The Indian mackerel is the target species of coastal fisheries of many countries around the Indian Ocean rim and Southeast Asia. While large purse seines are used in some areas, artisanal fisheries using lift nets and gill nets account for high proportions of the total catches. It is estimated that the annual catch of all three *Rastrelliger* species would be at least 250 000 tonnes.

BUTTERFLY MACKEREL

Gasterochisma melampus

OTHER COMMON NAMES
	Butterfly kingfish
	Butterfly tuna
	Scaly mackerel
France	Thon papillon
Japan	Urokomaguro
Spain	Atún chauchera

An unusual looking scombrid, the butterfly mackerel has an archaic, almost primitive appearance. Indeed, this unusual species is thought to physically resemble early, or ancestral tunas in at least some of its features. It is an outlier within the family Scombridae, being the only member of its own subfamily, Gasterochismatinae, whereas the other scombrid subfamily, Scombrinae, contains all of the other 48 species of tunas and mackerels.

Although primitive in general body form, the butterfly mackerel has some internal features that are considered to be quite advanced. Very little is known about its overall biology, but anatomical studies have shown that the butterfly mackerel possesses the largest brain/eye heating organ of any fish. This organ, also well developed in broadbill swordfish (but evolved independently), is a highly modified eye muscle situated near the brain at the rear of the eyeball. However, instead of contracting, the mass of tissue actively produces heat which keeps the vital brain and eye functioning efficiently, especially in cold water. It is thought that this extraordinary organ gives its owner a huge advantage over more sluggish prey species of fish and squid. The names butterfly mackerel or butterfly tuna seem misplaced, until it is realized that juvenile fish are equipped with enormous pelvic fins which do resemble the wings of a butterfly (or perhaps a moth). However, why this is so is completely unknown, and the pelvic fins assume more normal proportions as the fish grows toward adulthood.

There are two features which set the butterfly mackerel apart from all the other tunas and mackerels. Firstly, its entire body is covered by large, round (cycloid) scales – the only scombrid with such scalation – and secondly, the pelvic fins, which are enormous in juvenile fish, fold neatly into a ventral groove, whereas

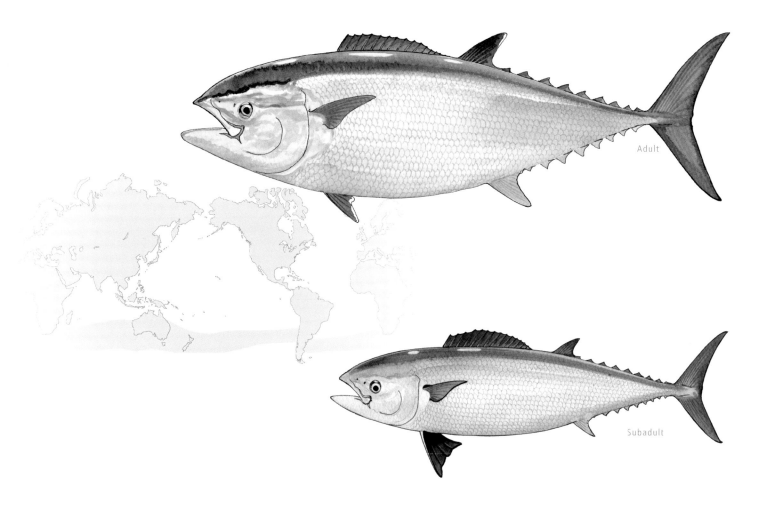

Adult

Subadult

these fins either fit into a depression or fold flat against the body in the other tunas and mackerels. Butterfly mackerel are normally dark blue to black on the dorsal (top) surface, grading to a silvery mirror-like sheen on the flanks and belly. The whole body fades to dark gray on death.

The butterfly mackerel is a circum-global species occurring in a band right around the Southern Ocean, mainly between latitudes of about 35°S to 50°S, but extending as far as about 27°S adjacent to southern land masses. No similar or equivalent species occurs in the northern hemisphere. Interestingly, a single specimen of the butterfly mackerel was reported north of the Hawaiian archipelago, so it is likely that, on occasions, some vagrant individuals wander far from the species' usual range. Butterfly mackerel appear to have clearly defined temperature preferences, being largely confined to waters within the narrow range of 8°C to 10°C.

Nothing is known about the movements of butterfly mackerel, other than that there is an apparent general expansion of the species' range to the north

during the southern (austral) winter, and contraction southward in the summer.

Butterfly mackerel grow to at least 164 cm in length, which would equate with a weight of about 80 kg (175 lb). It appears that the largest specimens are found on the western sides of all three major oceans compared with the eastern areas. This suggests some structure of populations, but no genetic work has been carried out on the species.

Because of its cold-water distribution, the butterfly mackerel is virtually never caught by recreational fishers, and therefore IGFA does not include the species in its list of gamefish species.

The butterfly mackerel shares the waters with, and is almost exclusively captured in association with southern bluefin tuna, especially by longline fleets which fish well into the Southern Ocean for the lucrative southern bluefin. Even though the butterfly mackerel grows to a relatively large size, and is sometimes prolific, it has relatively low value in its own right, due apparently to its extremely oily flesh, and is therefore nearly always discarded at sea in unknown, but possibly large quantities.

Mackerels

MACKERELS

WAHOO
Acanthocybium solandri

OTHER COMMON NAMES

France	Thazard-batard
	Ono
Japan	Kamasu-sawara
Spain	Peto

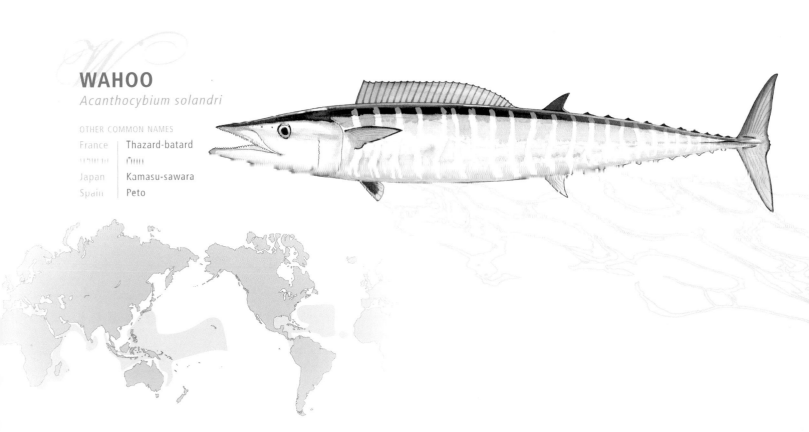

The name 'wahoo' sounds like it might have been coined (and uttered) by the first person who hooked one on a handline. This is an incredibly fast-swimming fish, perhaps the fastest of any fish in the ocean.

The real origins of the name are a little obscure, with perhaps the best explanation being the fact that early European explorers may have first come across the species in the Hawaiian islands, the main island of which was, and still is, called Oahu (pronounced 'o-wa-hoo'). In fact, old maps of Hawaii often used the spelling 'wahoo' for this island. The Hawaiians themselves appear not to have used this name, however, since the local name for wahoo is 'ono', which means 'good to eat' in the native tongue. Another derivation is suggested as coming from the North American Indian word 'wahnhu' meaning 'burning bush'.

The wahoo is a member of the tribe Scomberomorini, within the family Scombridae, which includes all of the tunas, mackerels and related species. The Scomberomorini consists of 19 species, 18 of which are Spanish mackerels, or seerfishes, in the single genus *Scomberomorus*. Within this tribe, therefore, the wahoo sits by itself, the only member of the genus *Acanthocybium*.

Several features immediately distinguish the wahoo from the Spanish mackerels. Firstly the wahoo has a markedly elongated, pointed snout, giving the appearance of its eye being set well back, as well as a 'long-nosed' appearance (the anterior eye margin is about equidistant from the tip of the snout and the rear of the operculum); secondly, its tail is small and relatively vertical compared with the larger, sweeping tails of the mackerels; and thirdly, all of the mackerels have gill rakers (comb-like projections on the front of the gill arches) while wahoo completely lack them. Lastly, the wahoo, as a single and very successful species distributed across the whole globe, is a true 'cosmopolitan' species. In contrast, the 18 species of mackerels are much more restricted in their ranges, all occurring near land masses, some having a total distribution of only a few hundred thousand square miles.

Although the wahoo is oceanic and widespread, it has a somewhat disjunct distribution in all three major oceans. It is a tropical to subtropical fish, occurring near the major continents, mostly wide of continental shelves. The concentration of wahoo in each of the major oceans tends to be highest in the western side of each.

Virtually nothing is known about movements of wahoo, with very few ever having been tagged and released. Given its broad oceanic distribution and tendency to appear seasonally, it is likely that individuals are capable of long-distance migrations across open water. However, until extensive tagging experiments are undertaken, postulated long-distance movements are quite speculative.

Wahoo grow to impressive sizes. The maximum size quoted in the scientific literature is 210 cm and 83 kg (182.6 lb) and the all-tackle world angling record recently confirmed this with an 83.5 kg (183.7 lb) fish caught off Cabo San Lucas, Mexico, in 2005. The Atlantic record is considerably smaller – 70.5 kg (155 lb), caught off the Bahamas in 1990.

Very small wahoo, less than about 45 cm long, are rarely seen, possibly because the species has a very rapid growth rate. Indeed, one wahoo which was tagged in the western Atlantic was recaptured after just ten months, during which time it had grown from about 5 kg (11 lb) to 15 kg (33 lb), while another recent study suggests that wahoo may even reach 13.6 kg (30 lb) by their first birthday.

Wahoo may have extended spawning

A pod of wahoo
hover in formation
beneath a fishing
boat.
David Itano

seasons since fish in various stages of reproductive activity are often found in the same areas. A single specimen of 131 cm in length was estimated to contain about 6 million eggs – a high number for a fish of this size. As well as a rapid growth rate, wahoo probably mature at quite small sizes. A study of wahoo caught both inside and outside the Gulf of Mexico showed that some males as small as 3.2 kg (7 lb) and females as small as 5.5 kg (12 lb) were sexually mature.

Both the upper and lower jaws of wahoo are equipped with a single row of extremely sharp, finely serrated teeth which they use to slice through prey species with a single motion. Their chopping action is made even more efficient by the fact that they are one of few species of fish which have a movable upper jaw, increasing their gape and allowing them to cut right through quite large prey. It is not uncommon for wahoo to attack trolled baits by slicing them cleanly just behind the hook. Wahoo will attack most pelagic prey, including tunas, flying fishes, scads, pilchards and squid. One interesting aspect of their biology is that,

wherever they are found throughout the world, their stomachs nearly always contain two (sometimes three) large parasitic worms over 2 cm in length. This worm, a digenetic trematode, *Hirudinella ventricosa*, is highly host-specific, meaning that it only occurs in wahoo.

For many of the truly open-ocean fishes, their offshore distribution makes their study difficult. This is especially true for the wahoo, since it is not a schooling species and is therefore rarely an important commercial proposition (although it is sometimes canned in American Samoa as a bycatch of tuna longlining). Nevertheless, it is an excellent and popular table fish, and is especially marketed in Hawaii under its local name, 'ono'. The demand for ono in Hawaii always outstrips local supply, to the extent that much of the ono on Hawaiian restaurant menus will have been imported from all over the Pacific. Wahoo are not often targeted by recreational anglers, although if it is known that large wahoo are present in an area, specialized lures rigged with strong wire traces may be put out to catch this welcome addition to the day's fishing.

Wahoo and
dolphinfish under
a floating log.
Both species are
strongly attracted
to floating
objects.

SPANISH MACKERELS
Scomberomorus spp.

The Spanish mackerels (also called seer-fishes) all belong to the genus *Scomberomorus*. They are members of the family Scombridae and, together with the single species, the wahoo (*Acanthocybium solandri*), form the tribe Scomberomorini. Spanish mackerels can be distinguished from the similar looking, but very different wahoo by the length of the snout. Measured from the front of the eye, the snout of the wahoo is as long as the rest of the head, while the snout of the mackerels is much shorter. Also, a quick check will show that mackerels have a few gill rakers while wahoo have none.

All of the *Scomberomorus* species have long, very sharp, knife-like teeth as well two small lateral keels on the caudal peduncle. They also lack an obvious corslet of scales around the head and shoulders, a feature common among all of the tunas.

There are 18 species of Spanish mackerel. All are streamlined, pelagic predators which primarily occur in tropical to subtropical neritic habitats; that is, in extensive shallow to medium depth coastal areas often extending offshore. As well, some species occur around islands, reefs and atolls. The six mackerels chosen here are either the most widespread, or species that would be of particular interest to the targeted audiences of this book.

Narrow-barred Spanish mackerel
Scomberomorus commerson

The narrow-barred Spanish mackerel is an extremely important species throughout its considerable range. While many of the mackerels have quite restricted distributions, the narrow-barred Spanish mackerel proves that rules always have exceptions. This species has a continuous distribution from the Cape of Good Hope, along the eastern African coast to the Red Sea and Persian Gulf, around the coasts of India and Sri Lanka and Southeast Asia to Japan, throughout the Indonesian archipelago, and through to Australia where it is quite common above the Tropic of Capricorn, extending to a latitude of about 30°S on both the east and west coasts. Good populations are also found around New Caledonia and Fiji in the western Pacific.

The lateral line of the narrow-barred Spanish mackerel dips downwards behind the level of the second dorsal fin. The body is silvery gray and the flanks normally prominently barred with many vertical, wavy lines, sometimes breaking into spots ventrally. These bars are often very distinct, but can also be quite faint, especially after death. The first dorsal fin of Spanish mackerel is a uniform dark blue-black.

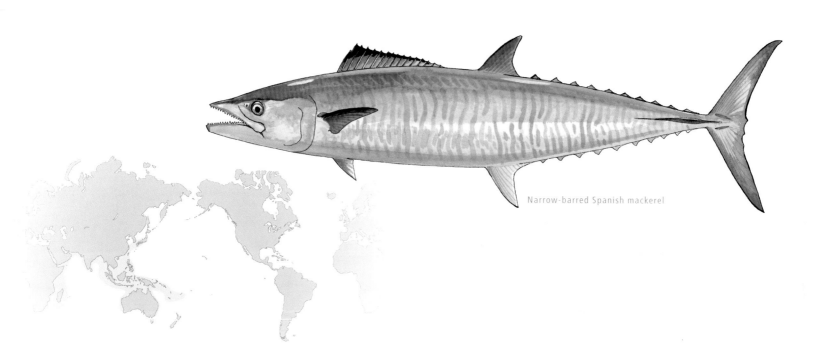

Narrow-barred Spanish mackerel

Through conventional tagging, narrow-barred Spanish mackerel have been shown to make extensive movements along the east coast of Australia. Some fish have moved coastally over distances in excess of 1800 km. On the other hand, tagging has also shown that some fish appear to be resident around certain reefs, simply scattering after the spawning season and returning the following year. There is good evidence from genetic studies that there are at least two separate stocks of narrow-barred Spanish mackerel in northern Australia/Papua New Guinea. This implies that there may also be other discrete stocks of the species throughout its extensive range, but the genetic studies to determine this are yet to be undertaken.

Like many other pelagic fishes, narrow-barred Spanish mackerel grow very quickly in their first few months. Juveniles measuring up to 40 cm are estimated to be only about three or four months old. They reach about 60 cm (2.2 kg – 5 lb) by their first birthday, about 5 kg (11 lb), or 80 cm (for females) by their second and about 9 kg (20 lb) or 90 to 100 cm by their third. The official IGFA world record for the narrow-barred Spanish mackerel is 44.9 kg (98.8 lb) for a fish landed off South Africa in 1982. Long-standing records such as this usually indicate that the record weight is towards the upper limit attained by a species. However, in the case of the narrow-barred

Spanish mackerel, the largest size quoted in the scientific literature is 70 kg (154 lb) – much larger than the world record of 44.9 kg. This was a fish caught by a commercial fisherman off northern Queensland, and its weight and length (240 cm) were verified from a sales docket and photograph. While there may have been some error in these figures, this must still have been an extremely large mackerel.

Narrow-barred Spanish mackerel spawn at different times depending on latitude. Large females probably produce tens of millions of eggs, and spawn several times during each spawning season. Little is known about the earliest life stages of the species until they reach a size of about 10 cm, when they appear in the shallows, even sometimes being found in tidal inlets and estuaries.

Narrow-barred Spanish mackerel occasionally undertake an amazing hunting strategy. Coming in from the side of a trolled bait or lure, the fish will suddenly leap clear of the water at speed in a low, horizontal trajectory, invariably coming down right on target. Presumably, this tactic allows the predator to completely surprise its prey by apparently appearing out of nowhere. The method of attack certainly indicates excellent eyesight at a distance, as well as an incredible degree of anticipation and coordination.

Throughout Southeast Asia, right around the Indian subcontinent and along the east African coast, the narrow-

Above Narrow-barred Spanish mackerel often travel in small groups or 'pods'. This species has the broadest geographical distribution of any of the Spanish mackerels.
seapics.com/ James D. Watt

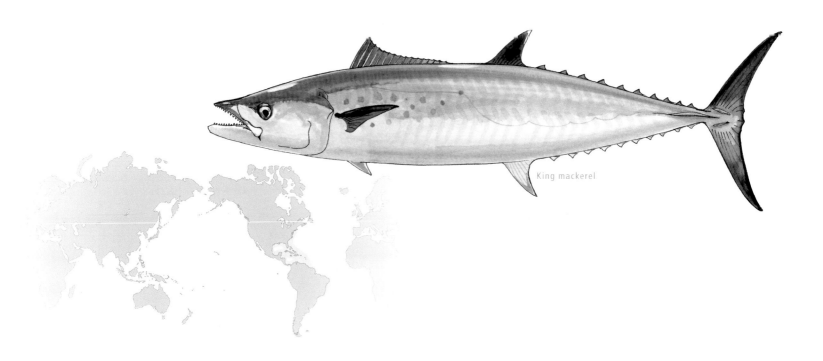

King mackerel

barred Spanish mackerel is the primary target of large fleets of coastal gill-netting boats. Although the gear is set and hauled by hand, it is still effective in taking large catches of this prized fish. Unfortunately, nothing is known regarding stocks of narrow-barred Spanish mackerel in these widely spread developing regions.

King mackerel
Scomberomorus cavalla

The king mackerel (or 'kingfish') is the largest of the Spanish mackerels that occur in the western Atlantic. It has an extensive distribution along the eastern seaboard of the Americas, from Massachusetts to Brazil.

Coloration of the king mackerel is relatively plain, without the prominent yellow-orange spots or streaks which are characteristic of the Spanish and cero mackerel respectively. It is important to note, however, that king mackerel may have dull bronze colored spots towards the front of the body, especially when small. The other definitive feature of the king mackerel is its lateral line, which curves down markedly at the level of the second dorsal fin.

The king mackerel reaches a con-siderably larger size than the cero or the Spanish. While the latter species may reach 6 kg (13 lb) and 7.5 kg (16.5 lb) respectively, the king is known to attain at least 42.18 kg (93 lb) – the current IGFA record – and specimens over 10 kg (22 lb) are not uncommon. Growth rates of king mackerel are somewhat uncertain, but the species is thought to be relatively long-lived. In one study, the oldest female fish aged was estimated at 14 years old (140 cm from snout to tail fork), while the oldest male aged at 12 years and measured 98 cm.

Surprisingly, maturity may be reached for females at only one year of age, corresponding to a length of just 35 cm. As with other mackerels, males mature later, at four years old, at which age they would be about 75 cm long. All fish over about 7 kg (15 lb) would be most likely females. It has been estimated that a female fish weighing 25 kg (55 lb) would produce 12.2 million eggs.

The king mackerel is enormously important, both commercially and rec-reationally. There have been concerns regarding stocks in the 1990s, but the current consensus is that stocks are not being overfished.

Spanish mackerel
Scomberomorus maculatus

Many of the mackerels have the name 'Spanish' in their title, but this species is the only one to be given the single moniker as its official common name. It is found only along the eastern seaboard of the United States, extending to the Yuca tan Peninsula.

Genetic studies suggest that there are possibly as many as six separate populations of Spanish mackerel that may mix on feeding grounds off Florida, but which probably have separate spawning grounds.

The Spanish mackerel is most likely to be confused with the king and cero mackerels which occur in the same region. It can be separated from the king mackerel by the presence of yellow-orange spots on the dorsal surface, and a lateral line that does not dip sharply at the level of the second dorsal fin (as does that of the king mackerel). Body coloration also separates the Spanish mackerel from the cero. While sharing yellow-orange spots, the cero has a distinctive orange-yellow stripe, sometimes broken, running along the midline of the body, which the Spanish mackerel lacks.

Spanish mackerel are thought to live for up to eight years and grow to a size of about 6 kg (13 lb). All fish larger than about 50 cm (fork length) will be female. The species matures at a very young age. Females measuring only 25 to 35 cm, and only one year old may be mature, and males at an even smaller size. As for other scombrids, Spanish mackerel are highly fecund, with the larger females (over 60 cm long) producing more than one million eggs each spawning season.

As well as being a prime commercial species, the cero mackerel is also an important recreational sportfish. Surveys show that some two million are taken annually off Florida alone. Interestingly, nearly 10 per cent of that catch is taken offshore, between 3 and 200 miles.

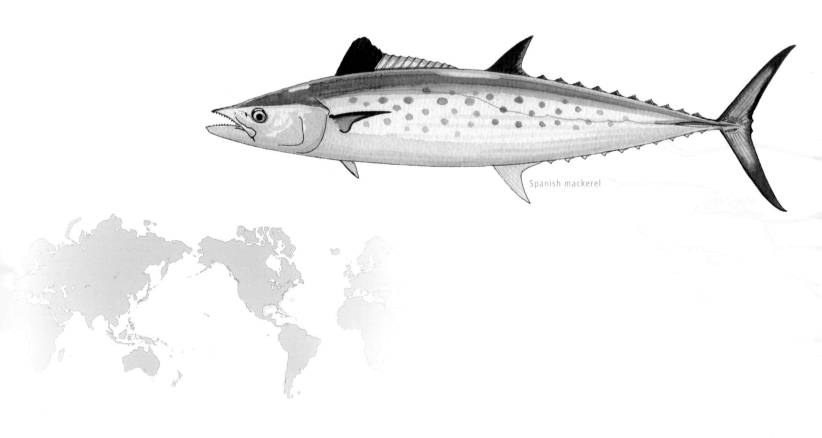

Spanish mackerel

Cero mackerel
Scomberomorus regalis

The cero mackerel is the third of the western Atlantic Spanish mackerels treated here. As noted, it occurs in the same general region as the king and Spanish mackerels, with its geographic range extending from Maine to northern Brazil. It has a patchy distribution in the Gulf of Mexico but is relatively common in the Caribbean.

As also noted above, the cero mackerel may be confused with both the Spanish and king mackerels. The cero, though, is immediately identified by the longitudinal yellow-orange stripe (sometimes broken) which runs down the midline of the body (with small spots on either side of the line), a feature which the other two mackerels lack.

The cero is a relatively small mackerel, growing to a maximum weight of 7.7 kg (17 lb) – the IGFA all-tackle record caught off Florida in 1986. Oddly, a maximum length for the species of 183 cm is oft-repeated in the literature, but this must surely be in error since a mackerel of that length would weigh at least 30 kg (66 lb).

As with other mackerels, fish mature at a relatively small size – about 33 cm for males and 38 cm for females. Growth rates do not appear to have been studied for this species.

While cero mackerel are not caught in the quantities that Spanish and king mackerel are, it is still a very popular species for sportfish anglers throughout its range.

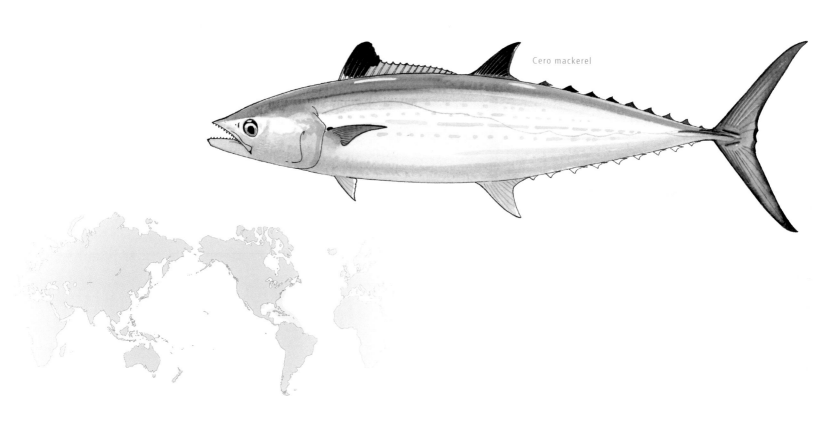

Cero mackerel

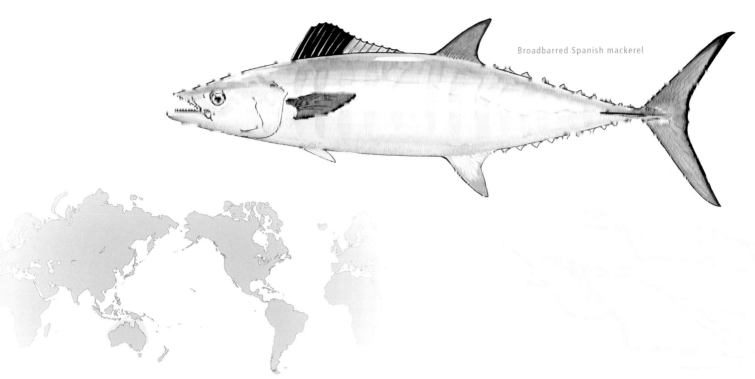

Broadbarred Spanish mackerel

Broadbarred Spanish mackerel
Scomberomorus semifasciatus

The broadbarred Spanish mackerel (or broadbarred king mackerel) occurs only around the northern half of Australia, extending to southern Papua New Guinea. Freshly caught fish are marked with 12 to 20 (or so) broad vertical bars on the upper part of the body, not to be confused with the narrow-barred Spanish mackerel, adults of which have 40 to 50 vertical bars covering the entire flanks. (Juvenile narrow-barred mackerel may have less than 20 bars, but those markings are vivid and remain after death.) The bars of the broadbarred mackerel fade quickly after death, becoming indistinct blotches, or disappearing altogether. The fish loses its silvery sheen, leading to its other quite appropriate name, the gray mackerel.

Growth rates of broadbarred mackerel are quite rapid for the first few years of life, attaining lengths of about 75 cm for females and 70 cm for males by the end of their third year. Broadbarred mackerel are estimated to be quite long lived compared with other small mackerels, with both males and females apparently reaching ages of 11 to 12 years (however, the bulk of the population would be aged eight years or less). As is the case for all species of Spanish mackerel so far studied, females apparently grow faster than males, although the discrepancy in growth rates, and maximum sizes of the sexes is not as great as for other species.

Broadbarred Spanish mackerel spawn over quite an extended period, from September through January. Again, this contrasts somewhat with several other small mackerels, which have marked peaks in spawning of relatively short duration. During an intensive tagging study, only small numbers of broadbarred mackerel (313) were tagged, a result which indicates either that the species is difficult to catch on hook and line, or that its numbers were low in areas of tagging. The former reason may well be the case since commercial catch statistics show that catches of this species are quite large (400 to 600 tonnes per year – second only to narrow-barred mackerel) while the recreational catch for the same area was estimated at only 12 tonnes. Broadbarred mackerel are targeted commercially with nets, not trolled lines, so it does appear that this species does not readily take lures or baits, indicating different feeding habits compared with other mackerels.

Australian spotted mackerel
Scomberomorus munroi

The Australian spotted mackerel may be distinguished from its closest look-alike, the school mackerel (*Scomberomorus queenslandicus*) by the size of its body spots. These are less than the diameter of the eye in spotted mackerel and larger than the eye diameter in school mackerel. In both cases, spots can be fairly indistinct, but if anything, are usually more obvious in spotted mackerel. Another handy distinguishing feature is the color of the first (folding) dorsal fin. In spotted mackerel, this fin is dark blue/black over almost its entire surface, whereas the same fin in the school mackerel is jet black, but with a contrasting, intense white patch in the middle of the fin.

As noted, many of the mackerel species have quite limited distributions, but having said that, the spotted mackerel is somewhat more widespread than the other Australian small mackerels. Spotted mackerel are strictly native to Australia and Papua New Guinea, extending down the east coast to the central New South Wales coast in most summers, and in particularly warm years, even appearing as far south as Sydney and beyond in reasonable numbers in some years. The summer extension of spotted mackerel is quite predictable, and is the basis of a targeted recreational fishery for the species in late summer.

Australian spotted mackerel spawn in late winter/early spring, apparently stimulated by warming water, with the peak spawning month being September. Studies strongly suggest that the spotted mackerel is a fast-growing, short-lived species in which females grow faster than males and probably live longer. This means that, like most of the other mackerels, all the largest fish in a population will be female. A one-year-old fish will average about 50 cm, and two-year-olds, a little over 60 cm. The oldest female fish so far aged was a specimen estimated at five years old which measured 85 cm (fork length), and weighed about 6 kg (13 lb). Most males do not live beyond four years, at which age they would average a little under 70 cm in length, while four-year-old females would average a little over 80 cm.

Some sources in the literature state that spotted mackerel grow to a maximum of 100 cm in length and 8 kg (18 lb), or even 10 kg (22 lb) in weight. The Australian all-tackle record is 6.0 kg (13.2 lb), for a fish caught in 1996.

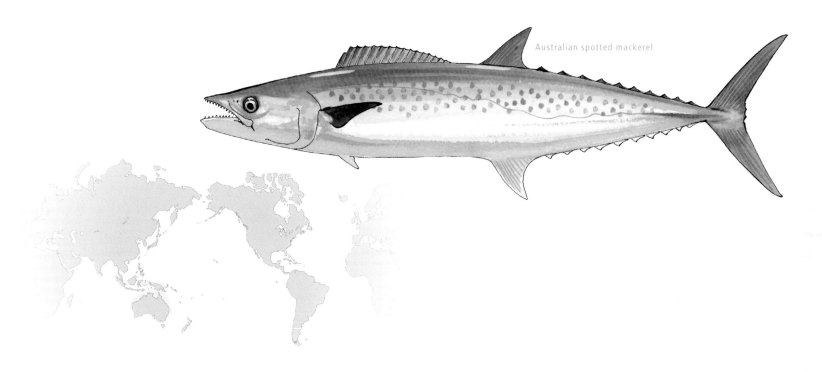

Australian spotted mackerel

SHARK MACKEREL,
DOUBLE-LINED MACKEREL

Grammatorcynus bicarinatus, G. bilineatus

Scaly tuna, Scaly mackerel *G. bicarinatus*

| France | Thazard requin |
| Japan | Nijosaba |

Scad *G. bilineatus*

| France | Thazard-kusara |
| Spain | Carite-cazón pintado |

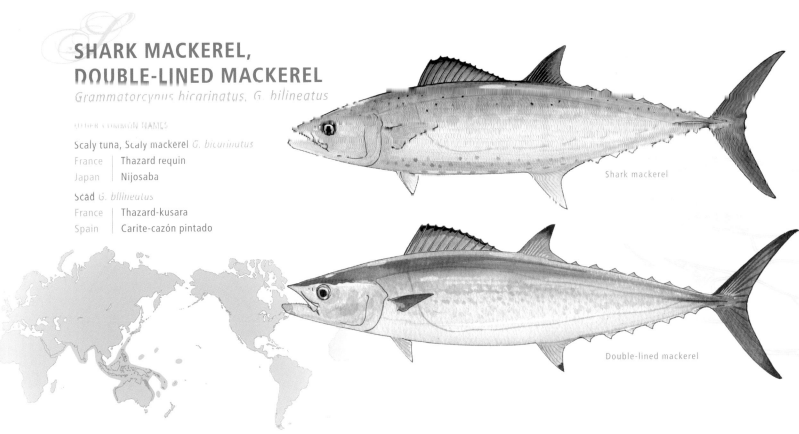

Shark mackerel

Double-lined mackerel

■ Double-lined mackerel
■ Shark mackerel

Shark mackerel are
readily identified
by their double
lateral line.
Julian Pepperell

These two members of the mackerel tribe (Scomberini) are closely related to each other, and form a separate branch in the tuna family, Scombridae. The flesh of both has a slight ammonia smell, not dissimilar to many shark species, hence the larger of the species being dubbed the 'shark mackerel'.

These distinctive fishes are easily identified by their prominent double lateral line, a feature that immediately separates them from the 'Spanish' mackerels, all of which have a single lateral line. These are relatively plain colored fish, the shark mackerel being mostly olive green on the back, fading to a pale belly, and the double-lined mackerel being dark blue on the back. Although shark mackerel grow to a much larger size than double-lined mackerel, small shark mackerel may often be confused with double-lined mackerel where their ranges overlap. In fact, for some time, they were thought to be the same species. Externally, shark mackerel are usually olive green on the back, while double-lined mackerel are usually blue. The other difference is a technical one – the relative size of the eye. In shark mackerel, the diameter of the eye is only about 3–4 per cent of the body

length, whereas in double-lined mackerel, it is larger (about 7–9 per cent of the body length).

The shark mackerel is native to the northern part of Australia, occurring on both the east and west coasts between latitudes of about 15°S and 30°S. Unlike other mackerels (*Scomberomorus* species) in the region, however, it does not occur through Torres Strait or around southern Papua New Guinea. The double-lined mackerel is much more widespread, occurring around the northern Indian Ocean, through Southeast Asia and around larger island groups in the southwestern Pacific.

The shark mackerel grows to about 13 kg (28.5 lb), while the double-lined mackerel only attains a maximum weight of 3 kg (7 lb). Most double-lined mackerel, however, are encountered at much smaller sizes, measuring about 30 cm and weighing about 0.5 kg (1 lb). Both species are keenly sought for use as baits for the black marlin charter fishery along the Great Barrier Reef in northern Australia. The double-lined mackerel is used for human consumption in other parts of its range.

JACKS/TREVALLIES

YELLOWTAIL
Seriola lalandi

The several members of the genus *Seriola* are similar in general features. The yellowtail is the most slender-bodied of the genus, its head being longer than the maximum body depth, a feature which separates it from the amberjack, *Seriola dumerili*. The yellowtail is also generally able to be identified since its tail is yellow, although juvenile samson fish, *Seriola hippos*, may also have yellowish tails. The second dorsal ray count easily separates yellowtail and samson fish, however – yellowtail have 31 to 34, while samson fish have 23 to 25. The other member of the genus, the almaco jack, *Seriola riviolana*, is identifiable by its relatively deep body, and more importantly, its distinctive high, pointed/hooked second dorsal fin. The maximum height of this fin is 1.3 to 1.6 times longer than the pectoral fin, whereas in the true amberjack, the species it most closely resembles, the height of the second dorsal is equal to, or only slightly longer than the length of the pectoral.

The yellowtail has a widespread distribution through the Indo-Pacific. It is common around New Zealand, the southern half of Australia, southern Africa and the southwestern United States and western Mexico. Although largely a coastal species, it is also found around some off-shore islands, including New Caledonia and the Galapagos Islands.

Yellowtail prefer water temperatures of 18°C to 21°C, but are also often found in cooler habitats. They tend to be confined to continental shelves, although large surface schools of juvenile yellowtail are sometimes encountered well out to sea. Interestingly, these offshore fish are reported to be very brightly colored, leading to some speculation about the possibility of the existence of another oceanic species, although this is most unlikely.

The most intensive studies of yellowtail growth rates have been undertaken in Australia. These estimate that yellowtail grow to a length of about 50 cm (fork length) after one year, 60 cm after two, 70 cm after three and 80 cm after four years. At that size, they would weigh about 8 kg (18 lb). This implies that yellowtail may be quite a long-lived species, since they grow to considerably larger sizes. The maximum size of yellowtail is a debatable point. While weights of up to 70 kg (154 lb) are sometimes quoted (especially in New Zealand and South Africa), the IGFA world angling record is 52 kg (114 lb), which is shared by two

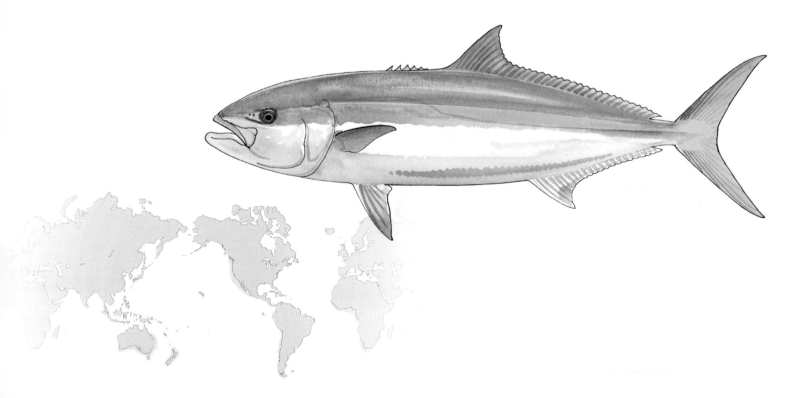

fish, both caught off New Zealand in the 1980s. The largest recorded yellowtail caught off California weighed 41.5 kg (91 lb), while the Australian record was a fraction smaller at 39.9 kg (88 lb).

Little is known about reproduction of yellowtail. The work which has been done suggests that they have a relatively long spawning period which varies with latitude. Off central Baja California, spawning occurs in mid to late summer, while in warm years, spawning has also been recorded off southern California. Off eastern Australia, spawning most likely occurs from August off central New South Wales, right through to February off southern New South Wales. Reports from anglers and commercial fishermen suggest that mass spawning events occur in surface waters well offshore, but more concrete evidence is needed to confirm this. Australian studies have found that male fish mature earlier than females, possibly as small as 36 cm long, with 50 per cent reaching maturity at a length of 47 cm. The size at first maturity for females is much larger, at 70 cm, or an age of about three years, with 50 per cent of females maturing by a size of about 83 cm. In New Zealand, females apparently mature at even larger, presumably older sizes, with 50 per cent of fish mature by the time they are 93 cm long and 100 per cent maturity reached by a size of 120 cm.

Virtually all of our knowledge on movements of yellowtail has been derived from recreational tagging of the species. In Australia, over 32 500 yellowtail have been tagged and more than 2100 recaptured. This recapture rate (6.5 per cent) is the highest of any species tagged on that program, and indicates a relatively high exploitation rate. Similarly, in New Zealand, about 12 000 kingfish have been tagged and about 1000 recaptured, again indicating a high recapture rate (8.3 per cent). The great majority of yellowtail tagged and recaptured have been juveniles (less than 60 cm long) and most have been recaptured within 50 km of their points of release, many near reefs or anchored buoys. The longest movements have been for three fish tagged off southeast Australia, two of which were recaptured off northern New Zealand, and one at Lord Howe Island. Demonstrating that movements across the Tasman Sea occur both ways, a New Zealand tagged yellowtail has also been recaptured off southeastern Australia. It should be stressed, though, that the long-distance movements are the exceptions rather than the rule. Nevertheless, these results do indicate that the yellowtail population of eastern Australia and New Zealand is all part of one common stock which mixes throughout its range. Genetic studies support this conclusion, but also show that crossings of the Pacific or Indian oceans probably do not occur.

Because of its predictable, schooling habits, the yellowtail has always been relatively easy to target both by recreational and commercial fishers. They have a strong tendency to school over reefs or wrecks, and also like to associate with floating objects such as buoys. This latter predilection, however, nearly saw the demise of the yellowtail off southeastern Australia. In the late 1980s, a new commercial fishing technique was developed, called yellowtail kingfish trapping. The large chicken-wire traps were not set on the bottom, but were buoyed to float at or near the surface over known yellowtail reefs. Yellowtail, being an incredibly inquisitive species, would enter the traps, even without the use of bait, and being gregarious, whole schools would follow, packing into the traps like sardines. Inevitably, numbers and sizes of yellowtail dwindled. After much debate, the traps were eventually banned by the state government in the late 1990s, and since then, the yellowtail has made a welcome comeback to its old haunts. Increased size limits have also very likely helped, and even commercial fishermen are reaping the benefits, with the average fish being larger and more valuable than before.

Juvenile yellowtail often form large schools which are easily targeted by various forms of fishing, including rod and reel and floating traps.
Phillip Colla/ oceanlight.com

AMBERJACK
Seriola dumerili

OTHER COMMON NAMES

	Greater amberjack
France	Sériole couronnée
Japan	Kanpachi
Spain	Medregal coronado
	Pez de limón

The amberjack is one of the largest members of the extensive trevally, or jack family, Carangidae. It is one of several members of the genus *Seriola*, all of which are elongated, robust fishes which grow to a large size and are known to occur in offshore waters, at least as juveniles. The amberjack is probably the best known of the Seriolas, and is a highly sought-after species by both commercial and recreational fishers.

The amberjack is similar in appearance to the yellowtail (*Seriola lalandi*), but with a more prominent, oblique black eyestripe running from the snout to the dorsal fin. In addition, the amberjack body is noticeably deeper than the yellowtail, while the tail of adult amberjack is dark gray to purplish brown in contrast, rather obviously, to the tail of the yellowtail which is yellow. One other useful identifying feature is the 'hinge' at the rear of the upper jaw (known as the supramaxilla). This is quite narrow in the yellowtail, but considerably broader in the amberjack. Where they overlap, amberjack and samson fish are often difficult to tell apart. The best identifying character is the fin ray count of the second dorsal fin. The amberjack has at least 29 of these fin rays, while the samson fish has between 23 and 25.

The amberjack is found in all three major oceans, and is not only confined to continental land masses like its relative the yellowtail but is also found around remote islands in the western and central Pacific, including a good population around the Hawaiian islands. Very young amberjack are often found well offshore associated with floating sargassum and it is likely that passive dispersal over considerable distances occurs during this phase.

The amberjack grows to the largest size of all *Seriola* species, with the current world angling record standing at 70.6 kg (155 lb). That weight was attained by two fish, both caught near Bermuda, one in 1981 and one in 1992. A larger specimen of 80.6 kg (177 lb) is also cited in the literature, but unfortunately, not verified. Interestingly, even though it has a worldwide distribution, all of the line class records for amberjack have been caught either at Bermuda, the Bahamas or between Virginia and Florida on the eastern US seaboard.

Although the biology of the species has not been studied in great detail, some work in the western Atlantic indicates that growth is quite rapid, with fish attaining between 37 cm and 50 cm by

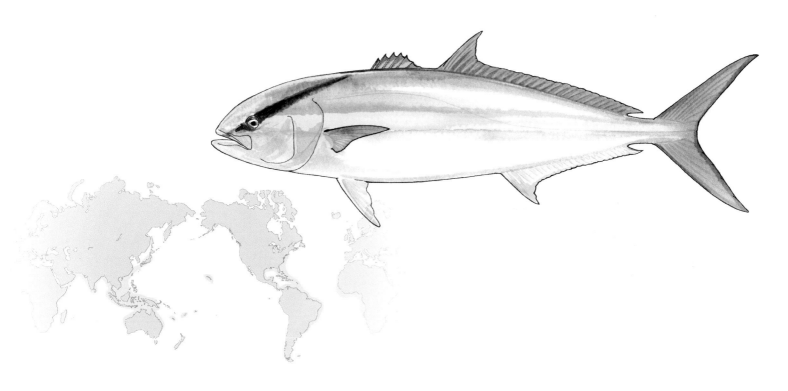

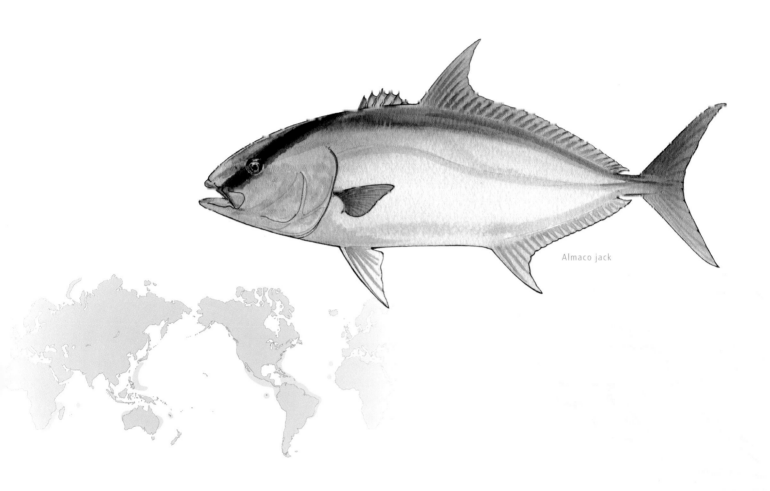

Almaco jack

the end of their first year of life, and by an estimated ten years of age, between 100 cm and 157 cm in length (over 5 feet). Maximum life span is estimated at about 17 years. Interestingly, female amberjack grow faster than males, and reach a larger size. As indicated in accounts of other species in this book, this type of dichotomy in maximum sizes between sexes occurs in quite a number of other species of pelagic fish.

Almaco jack
Seriola rivoliana

For completeness, mention should be made here of another prominent *Seriola* species, the almaco jack, *Seriola rivoliana*, also known as the highfin amberjack. This species is not particularly well recognized, and is quite probably commonly mistaken for greater amberjack when caught in the areas where both occur. It is a distinct species, however, identifiable by its relatively deep body, and more importantly, its distinctive high, pointed/ hooked second dorsal fin. The maximum height of this fin is 1.3 to 1.6 times longer than the pectoral (side) fin, whereas in the greater amberjack, the species it most closely resembles, the height of the dorsal is equal to, or only slightly longer than the length of the pectoral. The almaco jack is largely a tropical species, which occurs in all three major oceans, usually well offshore. It is particularly prized in Japan. While almaco jacks are usually encountered at small sizes not exceeding 10 kg (22 lb), they do grow to relatively large sizes. Most of the larger line class records listed by IGFA range between 22 kg (48 lb) and 28 kg (62 lb); however, the all-tackle record fish was much larger, weighing 59.87 kg (132 lb), caught off Baja California in 1964.

As noted, the almaco jack tends to be oceanic rather than associating with land masses, so if a *Seriola* species is caught on an offshore expedition to some remote reef or archipelago, it is always worth a close examination to determine its identity.

SAMSON FISH
Seriola hippos

OTHER COMMON NAMES
|| Sea kingfish
France | Sériole australienne

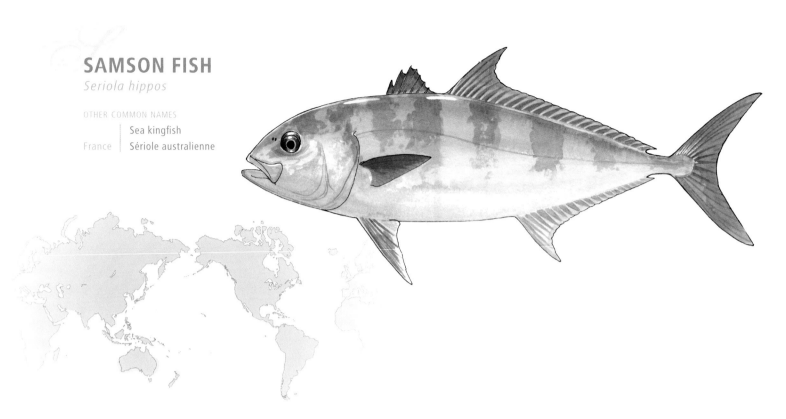

The samson fish is a member of the amberjack group within the jack/trevally family Carangidae. There has long been confusion over the identification of samson fish, *Seriola hippos*, and amberjack (*Seriola dumerili*) where their ranges overlap, and there are certainly marked similarities between these closely related fishes. One useful feature to look for is the unique reddish color of the teeth of the samson fish, apparently due to the gums usually being engorged with blood. Perhaps a more objective method to identify samson fish is to count the fin rays of the second dorsal fin. Samson fish have between 23 and 25, whereas the amberjack has at least 29, and for good measure, the other similar member of the group, the yellowtail (*Seriola lalandi*), has at least 31.

The samson fish is native to Australia and New Zealand, occurring nowhere else. In Australia, it has a patchy distribution on the east coast, but is far more common along much of the coast of Western Australia.

In a recent concentrated tagging study, a total of 8200 adult samson fish were tagged by recreational anglers at spawning aggregations off southwestern Australia. Results showed a rapid movement away from aggregation sites, presumably once spawning had occurred. For example, one fish tagged at the spawning site was recaptured 1100 km to the south after only 26 days, while a second fish moved 900 km south in 25 days. Two other tagged fish made unexpected journeys of over 2500 km eastwards across southern Australia, indicating a much more mobile population than previously thought.

Some preliminary work on growth rates has suggested that samson fish grow quite quickly when young – up to 60 cm by two years old and 80 cm by five years old. It is also apparent that individuals may well live for many years. The oldest fish aged was a 155 cm (total length) female, which was estimated to be 30 years old. It has also been noted that female fish grow to a larger size than males. Regarding the maximum size of the species, there is apparently a reliable record of a samson fish measuring 180 cm total length and weighing 53 kg (117 lb). This may be true, but the Game Fishing Association of Australia's (GFAA) all-tackle record stands at a much more modest 36.5 kg (80.5 lb), for a fish caught off southwest Australia in 1993.

Samson fish have been known for a number of years to form large, packed aggregations in summer, most probably for spawning, at which times they become particularly vulnerable to fishing. Several such aggregations occur annually between November and March off Rottnest Island in Western Australia, and researchers are now closely studying this phenomenon. One important question is: what proportion of the population of samson fish gather in these spawning aggregations? Indications from tagging suggest that quite high proportions might be involved, or at least, that many fish travel long distances to join in the spawning event. There are also strong indications that fish only spend a short time at the spawning aggregations before dispersing rapidly. Tagging has also indicated that fish return to the same spawning area, perhaps on an annual basis. One interesting finding to emerge is that, although the spawning aggregations of samsons occur in deep water, spawning does not seem to be confined to any particular area, or depth of water. Research on spawning fish also revealed another fascinating piece of information. Aggregating fish do not appear to be feeding while 'on site'. This was somewhat suspected by anglers who found that fish would only show interest in bait after being excited by flashing jigs. Just why this would induce otherwise disinterested fish to take lures, let alone baits, is yet another mystery.

Historically, spawning aggregations of samson fish seem to have been largely unknown. One of the eminent pioneers of fish biology in southern Australia, Edgar Waite, once described a school of samson fish on the surface, racing around his ship, sometimes within arm's length, and jumping clear of the water. In retrospect, he had probably encountered a spawning aggregation of this enigmatic species without realizing it.

Researcher tagging a samson fish as part of an intensive study of their movements
Andrew Rowland

BIGEYE TREVALLY
Caranx sexfasciatus

OTHER COMMON NAMES

	Great trevally
France	Carangue vorace
Japan	Gingame-aji
Spain	Jurel voráz

The bigeye trevally is a small to medium sized carangid which is identifiable by its fully scaled breast, white tips on its second dorsal and second anal fins (the latter sometimes indistinct), dark to black scutes (hard ridges on the tail wrist) and a small black spot on the upper rear of the gill cover. Not surprisingly, the bigeye trevally does indeed have a noticeably large eye. The one species which might be confused with the bigeye trevally is the gold spot trevally, *Carangoides fulvoguttatus*. However, these can be readily distinguished since, as noted, the breast of the bigeye is fully scaled while that of the gold spot is naked. Also, the second dorsal fin ray counts differ. The gold spot has 25 to 30, while the bigeye has 19 to 22.

The bigeye trevally has a broad distribution throughout the Indo-Pacific, including major island groups such as Micronesia, Hawaii and French Polynesia.

For many years, the bigeye trevally was commonly referred to as the largest of the group, and was even officially called the great trevally. However, it is now clear that the species grows no larger than a number of other medium-sized trevallies. Bigeye trevally reach a maximum size of about 80 cm fork length, possibly up to 1 metre. The IGFA record stands at 14.3 kg (31.5 lb), for a fish caught in the Seychelles, which is a fair indicator of its maximum size. Oddly, the official Australian record is 23.25 kg (51 lb), but this is likely to be a misidentification, since the next largest Australian specimen weighed 8.5 kg (19 lb) – far more consistent with sizes from other areas.

Bigeye trevally form large schools during the day which cruise along reef drop-offs in slowly swirling masses. Such schools are almost hypnotic in their slow movements and are often the subjects of underwater photographers looking for that definitive marine image. When night falls, however, like many other trevallies, the bigeye disperses and becomes a serious predator of the outer reef.

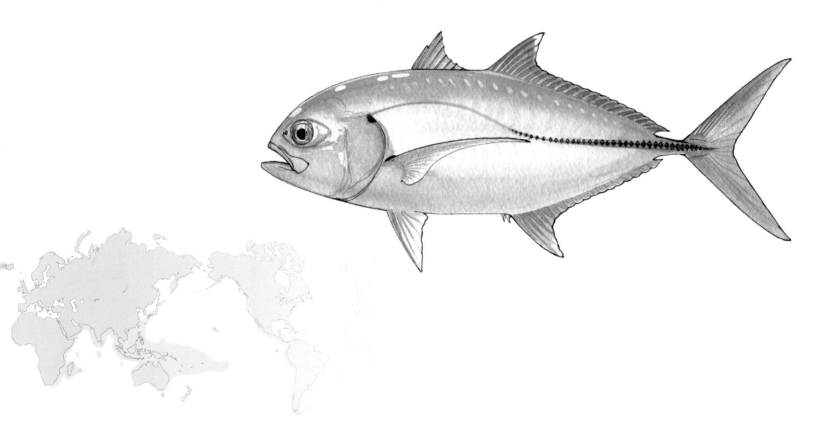

HORSE-EYE JACK
Caranx latus

OTHER COMMON NAMES

      Big-eye jack

      Horse-eye crevalle

France   Carangue mayole

Spain   Jurel ojón

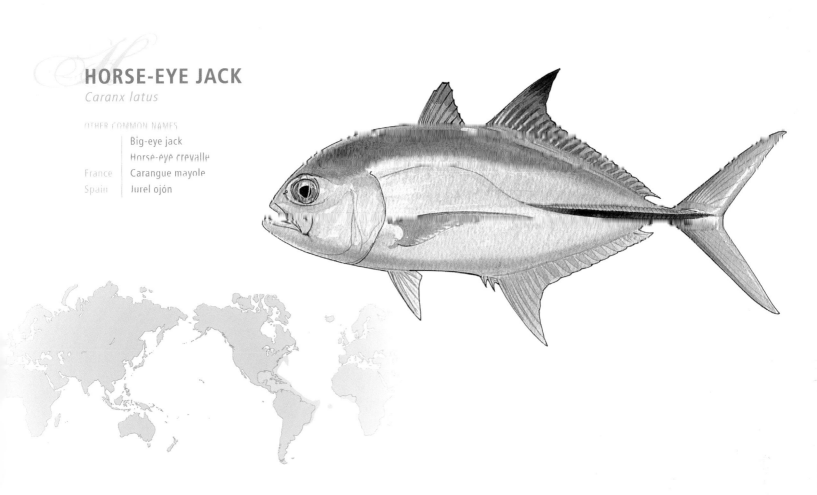

The horse-eye jack has a similar distribution in the western Atlantic to that of the bar jack – from New Jersey in the north to Brazil in the south. However, unlike the bar jack, the horse-eye jack is also found in the eastern Atlantic. This is one of the more pelagic, offshore species of jack which often swims in large slow-moving schools on the edges of deep drop-offs on coral reefs.

The horse-eye jack is somewhat elongated, but with a steeper head profile than the bar jack. Perhaps its best identifying combination of features is its yellow tail, which is usually very obvious, together with its silver body and contrasting dark scutes (sharp scales on the tail wrist). In looks and behavior, the horse-eye jack is very similar to the Indo-Pacific bigeye trevally, *Caranx sexfasciatus*. In fact, these two species would appear to be congeners, or counterparts in their respective oceans.

This is a medium-sized jack, growing to about 100 cm fork length, and a known weight of 13.4 kg (29.5 lb), which is the IGFA record, caught at Ascension Island in 1993.

A school of
bigeye trevally
is a mesmerizing
sight.
marinethemes.com/
Kelvin Aitken

GIANT TREVALLY
Caranx ignobilis

OTHER COMMON NAMES

	GT, Turrum
France	Carangue têtue
Japan	Ronin-aji
Spain	Jurel gigante

It is a bemusing fact that the scientific name bestowed on the giant trevally is *Caranx ignobilis*, which translates into the other common name once used for the species, the lowly trevally. Why this impressive fish was given such a dull name is anyone's guess, but it does seem obvious that the name was selected by someone who had never seen an adult giant trevally in full feeding mode. In fact, this is a very impressive fish.

The trevallies, or jacks, probably cause more identification problems for the layperson than any other group of fishes. Not only are there many closely related species, but body shape and coloration often change with the size and age of the fish. To make matters worse, some of the features which are used to identify trevally are not exactly 'user-friendly'. For example, the pattern of vomerine teeth, or the arrangement of scales on the breast area in front of the pectoral fin, are often key characters which are used to separate trevally species – all right in a lab but a little tricky on the deck of a boat.

As a case in point, identifying the giant trevally is not always easy, especially for smaller fish. The first feature to look for is the amount of scale-cover (termed 'squamation') on the breast area between the pectoral fins. In the giant trevally, this patch is naked (that is, has no scales), but importantly, right in the middle of this naked area is a small oval patch of scales. This is a good feature to identify the giant trevally, but, just to make matters a little difficult, another trevally, the brassy or Papuan trevally, *Caranx papuensis*, shares this feature and resembles the giant trevally in other ways. Fortunately, the Papuan trevally can also be separated from the giant trevally by its white trailing edge to the lower lobe of the tail fin, which the giant trevally lacks.

The giant trevally is distributed around the tropical and subtropical land masses of the Indian Ocean throughout Southeast Asia, including southern Japan, and around most of northern Australia. It is also found around isolated mid-ocean island groups, including Samoa, Hawaii

As the name suggests, the giant trevally is one of the largest of the speciose jack/trevally family.
James Watt/ imagequestmarine.com

and the Marquesas in the Pacific Ocean and Reunion in the Indian Ocean. It is around these islands that the largest giant trevally seem to occur, suggesting the formation of resident populations in these regions. Giant trevally are often found patrolling the edges of coral reefs and atolls where they sometimes congregate in their hundreds. Adult giant trevally are rarely found in the true open ocean, so it is likely that their dispersal to mid-ocean islands is achieved during larval or post-larval stages.

For some species of sport and game fish, there is a magic round weight, above which is regarded as a truly outstanding size to capture and/or release. For black and blue marlin and giant bluefin tuna, this number is 1000 lb (454 kg), while 100 kg (220 lb) is a magic number for yellowfin tuna. For giant trevally, the number is 100 lb (45.4 kg), and few anglers have attained that mark.

The all-tackle world record giant trevally weighed 66 kg (145 lb) and was caught off Maui, Hawaii, in 1991, while another weighing 59 kg (130 lb),

also caught off Hawaii, is the current world spearfishing record. Other large specimens have been recorded by game-fish anglers in widely separated locations – at Reunion Island in the Indian Ocean (54 kg – 119 lb), American Samoa in the western Pacific (53 kg – 116.5 lb) and Midway Island in the north central Pacific (47 kg – 103 lb).

While the official all-tackle record of 66 kg (145 lb) is certainly a very large fish, there are some reports of giant trevally growing even larger – to as much as 80 kg (176 lb). While that is possible, fish of this size are yet to be fully verified.

The giant trevally is a good example of a species which has very little, if any, commercial value, but which is of enormous potential value as a sportfish in many regions. The lure of the giant trevally is very strong, so much so that there is a large specialized market for fishing charters to remote locations to specifically target this single species. And as a bonus, in such fisheries, the great majority of fish are nearly always carefully released after capture.

CREVALLE JACK
Caranx hippos

OTHER COMMON NAMES

	Horse crevalle, Horse-eye jack
France	Carangue crevalle
Japan	Uma-aji
Spain	Jurel común

The crevalle jack is a solid, pugnacious-looking species of jack with a steep forehead and large eyes. It has a naked breast except for a small scaled patch just in front of the pelvic fins. Another distinctive feature is an oval black spot at the base of the pectoral fin.

It has a broad range in both the eastern and western Atlantic, In the west it is known from Nova Scotia in the north to Uruguay in the south, while in the east it occurs from Portugal to Angola and also inside the Mediterranean. It is also reported from the eastern Pacific, but it is most likely that those reports are in fact misidentified Pacific crevalle jack, *Caranx caninus*.

Juvenile crevalle jacks form schools, but fish tend to become solitary as they grow older and larger. They are not particularly important as commercial species, but are a highly sought sportfish.

Crevalle jack grow to more than 1 metre in length and a weight of over 20 kg (44 lb). The IGFA all-tackle record crevalle jack weighed 26.5 kg (58 lb) and was caught off Angola in 2000.

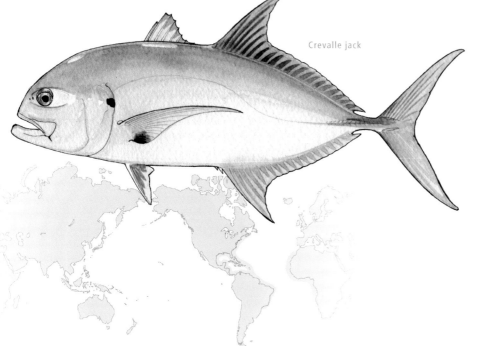

Crevalle jack

WHITE TREVALLY
Pseudocaranx dentex

OTHER COMMON NAMES

	White trevally
	Skipjack trevally
France	Carangue dentue
Japan	Shimaaji
Spain	Jurel dentón

The white trevally, also known widely as the silver trevally, is one of the most widespread of the Indo-Pacific trevallies. It is global in its distribution, occurring not only near continental land masses, but also commonly around islands. For example, it is a common species around New Zealand, and is also found at Lord Howe Island, Hawaii, the Azores, the Canary Islands, Madeira, Ascension and St Helena Islands. Not only are white trevally widespread geographically, they also occur throughout the water column, from the surface to the bottom. In fact, in some regions, schooling white trevally can be targeted by bottom trawl nets.

White trevally can be identified by their silvery-blue color, sometimes with a yellowish midline stripe, a black spot on the rear of the gill cover (operculum) and a scute count of 20 to 26 (scutes being the hard, sharp scales on the caudal peduncle or tail wrist).

White trevally are reported in an Hawaiian reference to attain a length of 122 cm and a weight of 18.1 kg (40 lb), although those sizes are not approached in many parts of the range, including Australia.

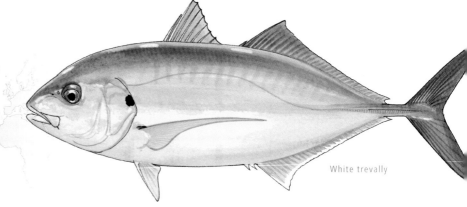

White trevally

GOLDEN TREVALLY
Gnathanodon speciosus

OTHER COMMON NAMES

France	Carangue royale
Japan	Koganeshima-aji
Spain	Jurel dorado

This is a distinctive species of trevally or jack, relegated to its own genus due to a number of features, the main one being its complete lack of teeth. Its coloration tends to be indeed golden, with its belly and sides flushed with a beautiful metallic yellow, usually broken by four or five vertical bars. Golden trevally also usually have a few dark blotches on their sides, although not all fish show this feature.

Juvenile golden trevally are bright yellow with five or six prominent black stripes, one through the eye, with narrower, less distinct stripes alternating between the broad ones. They may be regularly seen swimming near the heads of much larger fish, in much the same way as pilotfish, which they somewhat resemble, and perhaps mimic.

This is a widely distributed species throughout the Indo-Pacific. It occurs in a variety of habitats, both on inshore sand flats and on outer reefs. It is especially popular with saltwater fly-fishers.

The maximum size to which golden trevally grow is about 15 kg (33 lb).

A pair of golden trevally prowl around the edge of a bait school.
marinethemes.com/ Rob Torelli

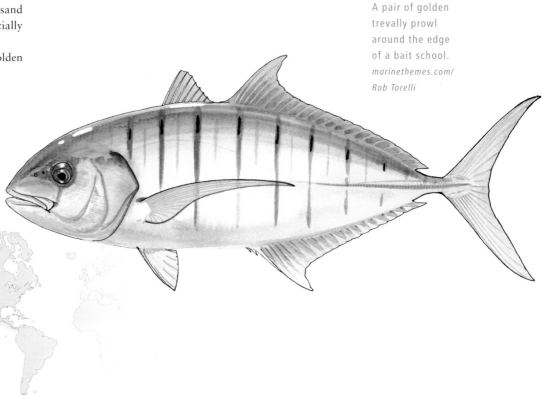

GOLDSPOT TREVALLY
Carangoides fulvoguttatus

The goldspot trevally is a widespread member of the carangid (jack) family, occurring throughout the tropical and subtropical Indo-Pacific (although not in the eastern Pacific). It occurs not only on continental shelves, but also around many islands, including Palau, New Caledonia and Tonga in the Pacific and the Seychelles and Reunion in the Indian Ocean.

It is identifiable by a scattering of small gold-colored spots mainly on the upper parts of body, a faint black spot on the gill cover (not always present) and a naked (scaleless) breast. Goldspot trevally often occur in similar habitats to the golden trevally (*Gnathodon speciosus*) and the giant trevally (*Caranx ignobilis*). Smaller fish are often seen in large schools, while, as is the case with many of the jacks, large adults tend to swim alone.

Goldspot trevally grow to about 120 cm long and a weight of 18 kg (40 lb). The species is not listed for record purposes by IGFA, but the Australian record is 12.6 kg (28 lb), caught in Western Australia in 1988.

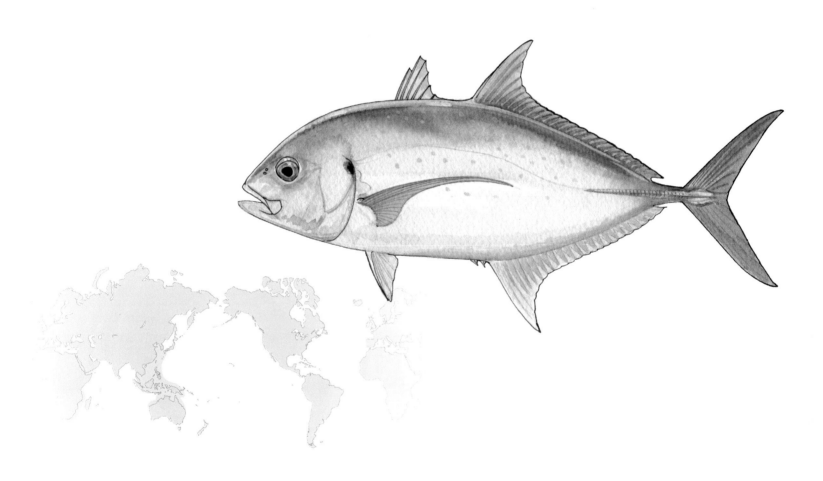

BLACK JACK
Caranx lugubris

OTHER COMMON NAMES

	Black trevally
France	Carangue noire
Japan	Kappore
Spain	Jurel negro

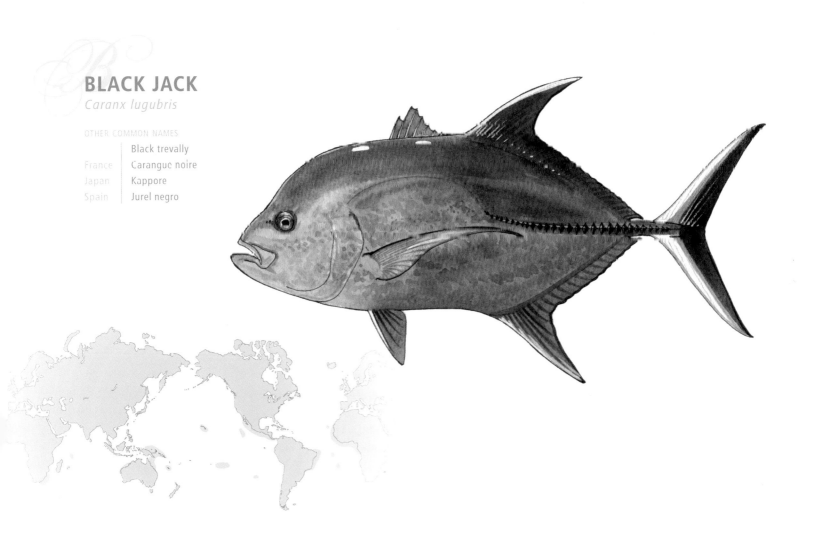

The black jack is a medium-sized worldwide member of the large jack/trevally family, Carangidae. It is largely a tropical species, occurring in all three major oceans. It has a particularly island-associated distribution, being found around many offshore island groups scattered throughout the tropics.

While many of the jacks and trevallies cause all sorts of problems with respect to correct identification, the black jack is a welcome exception to the rule. It shares a steep forehead with some other species, such as the giant trevally (*Caranx ignobilis*), but its color is quite distinctive, being dark gray to near black on the back and bluish gray on the flanks and belly. Importantly, the scutes (sharp scales on the tail wrist) are also black – a distinctive feature of this species.

The black jack grows to about 100 cm in length and a known weight of 17.9 kg (39.5 lb), which is the IGFA all-tackle record fish caught in the Revillagigedo Islands, Mexico, in 1995.

BAR JACK
Carangoides ruber

OTHER COMMON NAMES

	Red jack, Runner
	Neverbite
France	Carangue comade
Spain	Cojinua carbonera

This is a particularly attractive jack species that is found along the eastern seaboard of the United States, from New Jersey to Mexico and along the South American coast as far south as southern Brazil. The center of its distribution appears to be the West Indies region, where it is especially common. Bar jack prefer very clear marine waters around coral reefs and islands.

The shape of the bar jack helps to separate it from many of the other jacks and trevallies. It is a shallow-bodied fish with a markedly pointed snout. In life, its coloration is particularly striking, with a pale blue stripe high along its back and a pale leading edge to the lower lobe of the tail, which is otherwise dark or black. In contrast, the upper lobe of the tail is very pale.

The bar jack often swims in large schools, especially when young, but also when spawning. This is a relatively small species of jack, growing only to about 60 cm fork length, and a reported weight of 8.2 kg (18 lb), although the IGFA record for the species is just 3.5 kg (7.75 lb), caught off Miami, Florida, in 1999.

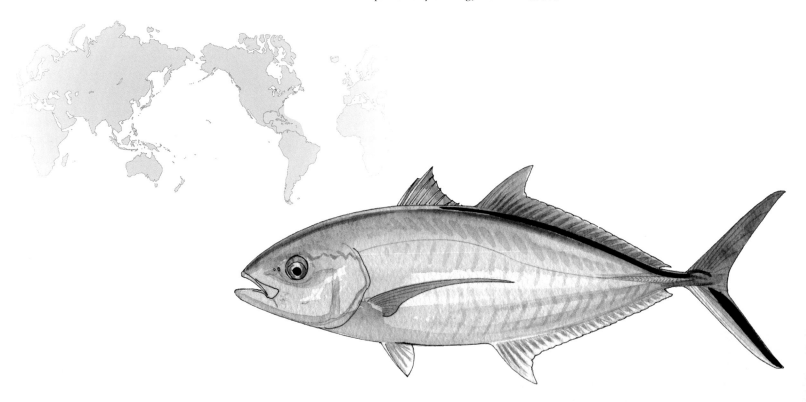

GREEN JACK
Caranx caballus

	Horse jack
France	Carangue royale
Japan	Koganeshima-aji
Spain	Jurel dorado

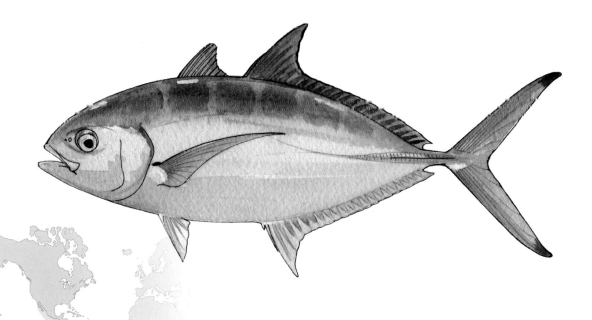

The green jack is rather elongated, with a fully scaled breast and an adipose (clear) eyelid covering the rear of the eye. As the name suggests, its overall coloration is greenish. This species is restricted to the eastern Pacific, where it occurs in large numbers. It is mainly found from southern California, along the Mexican and Central American coast, including the Sea of Cortez, to Peru. It also occurs at the Galapagos Islands.

Being particularly abundant, it is an especially important commercial and recreational species throughout its range as well as being a common food item of predatory fishes such as marlin and sharks. It may form large schools which travel into deeper offshore waters. It is estimated that green jack reach about 16 cm by year one and 35 cm by age three.

This is a relatively small species of jack, growing only to about 55 cm fork length, and a weight of about 3 kg (7 lb), although some reports suggest that it may reach 1 metre long.

SCADS
Decapterus spp., *Trachurus* spp.

The term 'scad' is a general name applied to many small members of the jack family Carangidae, especially to two distinct groups of schooling baitfishes. The 12 species in the genus *Decapterus* are mostly called scads, while the 14 species in the closely related genus *Trachurus* may also be called jack mackerels or horse mackerels – in fact, they are not mackerels (family Scombridae) at all, but therein lies the problem with the use of common names for fish. Both types of 'scad' are characterized by scutes (enlarged, hard scales) along the lateral line, with those of *Trachurus* species being particularly prominent along its whole length. All *Decapterus* species have a single small finlet behind the dorsal and anal fins, while *Trachurus* species lack this small fin.

Both types of scad are very important forage fishes for larger predators. They are also important baitfishes used by both commercial and sport fishers, especially as live bait for pole-and-line tuna fisheries. Some species – for example, the greenback horse mackerel, *Trachurus declivis* – are caught in large quantities and used for fishmeal products. The species selected here for brief description are both relatively oceanic and are typical of the two genera.

Mackerel scad
Decapterus macarellus

This is perhaps the most oceanic and widespread of the scads, occurring worldwide, not only on continental shelves but also around many islands in all three major oceans. This is the well-known opelu of Hawaii, a favorite bait for many forms of fishing. In contrast to many other small pelagic fishes, it has been observed, at least around Hawaii, that juvenile mackerel scad form large offshore surface schools, while inshore schools tend to be composed of larger adults.

Yellowtail horse mackerel
Trachurus novaezelandiae

The yellowtail horse mackerel is simply called the 'yellowtail', or more affectionately, the 'yakka', in Australia where it is an important baitfish. It occurs primarily over the continental shelf and, as is the case for many small pelagic species, juveniles mainly occur inshore while adults favor deeper, offshore waters.

It grows to a size of 50 cm, although average fish would be half that length. This is a surprisingly long-lived fish, with estimates of longevity put at 25 years.

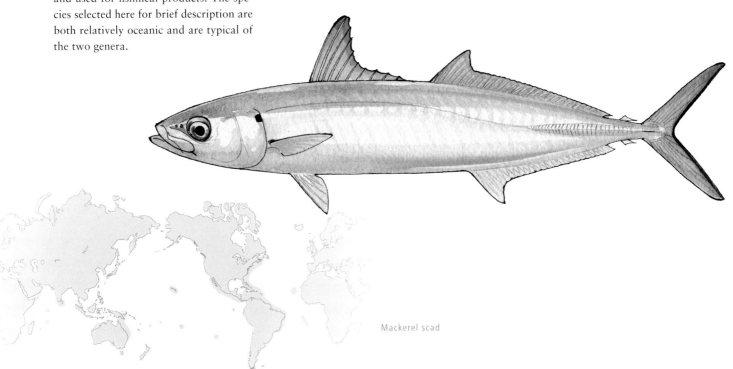

Mackerel scad

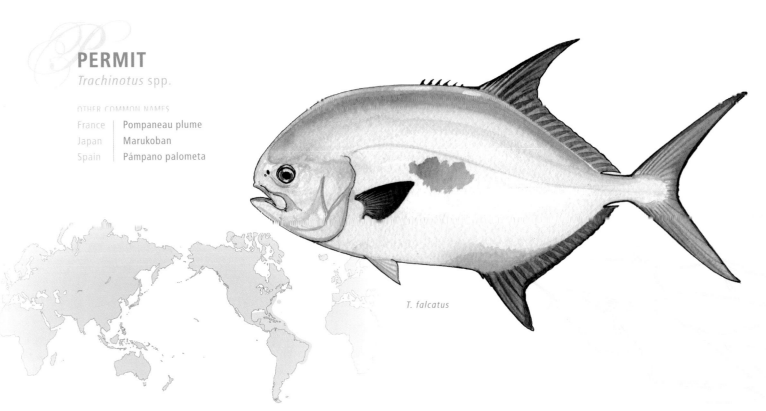

PERMIT
Trachinotus spp.

OTHER COMMON NAMES

France	Pompaneau plume
Japan	Marukoban
Spain	Pámpano palometa

T. falcatus

These highly distinctive looking fishes are in the pompano branch of the trevally or jack family, Carangidae. There has always been some confusion in different parts of the world over what constitutes a permit and what is a pompano. This is relatively easy to clear up since they are both, in effect, the same – that is, the permit is simply one particular species of pompano. Globally, there are at least 20 species of pompano, all belonging to the genus *Trachinotus*; however, only one of these has been officially dubbed with the name 'permit' – *Trachinotus falcatus*.

The 'true' permit is largely restricted to the western Atlantic, along the eastern American coasts, from Massachusetts to Brazil. It is particularly prevalent off eastern Florida and in the Bahamas and West Indies and has also been occasionally reported from the eastern Pacific. The permit is primarily a fish of the sand flats, and therefore, not a resident of the open ocean. However, it, and some of its close relatives, occur around oceanic islands and often remote atolls, indicating transport of individuals over some distances at some stage of the life cycle. Unabashedly, the permit is also included here because of its reputation as one of the most revered of marine sportfishes.

A closely related species, the Florida pompano, *Trachinotus carolinus*, is also found in the same region as the true permit and, although it rarely grows to a size much more than 3 kg (7 lb) – much smaller than the permit – separating the two species when small has always caused problems. Keen-eyed anglers could count the soft rays of the anal fin, which number 20 or less in the permit, compared with about 22 in the Florida pompano; however, a quicker method is to run one's finger over the tongue of the fish. The true permit has small teeth on the surface, while the pompano has a smooth tongue.

Species of *Trachinotus* occur in other parts of the world, where the name 'permit' is sometimes applied by sportfish anglers with some reverence. For example, there are four recognized species of *Trachinotus* in Australia, mostly called 'darts'. One species, *Trachinotus blochii*, the largest in the region, has long been known as the snub-nosed dart (also called 'oyster cracker'); however, in recent years, some writers and anglers have begun referring to this impressive species as the Indo-Pacific permit. One problem with using the name 'permit' is that, as noted, it has traditionally only

ever applied to one species of this largish group with all of the other members
of the genus *Trachinotus* being officially
called pompanos (although, just to confuse the issue, the fish called the 'African
pompano' is the only pompano which
is not in the genus *Trachinotus*, but is
rather a species of the threadfin trevally
genus *Alectis*).

Permit grow quite quickly, at least
for the first five years or so of life. Two-
to three-year-old fish would average 48
to 55 cm long. This is a long-lived fish,
estimated to live for perhaps 25 years or
more. Not surprisingly, the largest pompano in official record charts is indeed
the true permit, *Trachinotus falcatus*.
The current IGFA all-tackle record was

established in 2002 by a fish of 27.21 kg
(60 lb), caught off Brazil. A larger weight
for the species of 36 kg (79 lb) is quoted
frequently, but verification of this size has
proven elusive. Only one other species of
pompano is listed by IGFA – the African pompano. As indicated, this is not a
true pompano, but for completeness, the
record for that species (*Alectis ciliaris*) is
held by a fish weighing 22.9 kg (50.5 lb)
caught off Florida in 1990. Finally, one
other large species of pompano should
be mentioned, the southern pompano
of South Africa, *Trachinotus africanus*.
While the angling record for the species stands at 14 kg (31 lb), specimens
held captive in Durban Aquarium have
exceeded 25 kg (55 lb) in weight.

A pair of permit
prowl the sand
flats. There are
many species of
pompano, but at
least for now, only
one has the honor
of bearing the
name 'permit'.
*seapics.com/
Masa Ushioda*

QUEENFISH
Scomberoides spp.

OTHER COMMON NAMES

Leatherskin
Skinny fish
Queenie
France | Sauteur
Japan | Ikekatsuo
Spain | Jurel

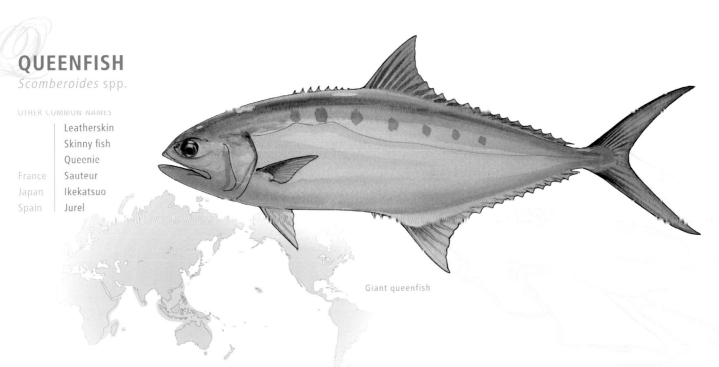

Giant queenfish

Queenfish are members of the Caringidae family, which also includes the jacks, trevallies, kingfishes and pompanos or darts. Closely related to the darts (also known as swallowtails), as well as the pompanos or permits, queenfish are also widely known by two other common names – leatherskins and skinny fish. Both names are appropriate, since larger specimens do indeed have very tough skin, and all species have highly compressed, or skinny bodies. ('Skinniness', by the way, is a body shape which a number of predatory fishes have evolved. It is thought that the thinness of the body, when viewed end on – that is, from the perspective of the prey – disguises the true size of the predator until it is too late.)

Queenfish can be readily identified by their extremely laterally compressed bodies, and the presence of several large oval markings along the flanks. All queenfish are closely related, being in the same genus, *Scomberoides*. There are four known species, all of which can be distinguished from pompanos and darts by their relatively short dorsal and anal fins, these fins being long and swept back in the latter two fishes.

Considering the four queenfish species, the giant queenfish or leatherskin, *Scomberoides commersonnianus*, is perhaps the easiest to identify. It has a blunt head profile, and a very large mouth, extending back well beyond the level of the rear margin of the eye. The flanks are marked with a row of five to eight nearly circular blotches positioned above the lateral line which are darker than the silvery green coloration of the body. The deep leatherskin or queenfish, *Scomberoides tala*, has a more pointed snout and elongated blotches, all of which cross the lateral line. The third species, the double spotted queenfish, *Scomberoides lysan*, is readily distinguished by two rows of small lateral blotches, one on each side of the lateral line. Finally, the needle-scaled or slender queenfish, *Scomberoides tol*, has a relatively short mouth, and small, faint blotches above the lateral line that are smaller than the eye.

All four queenfish species are primarily confined to the western Pacific and Indian oceans, with most species being concentrated through Southeast Asia and northern Australian waters. One species, the double spotted queenfish, is also found in Hawaiian waters. Most queenfishes are coastal in their distribution, with small fishes often entering bays and inlets, although muddy, turbid water is usually avoided. One species, the needle-scaled, is more oceanic, often occurring on outer reefs.

The largest of the queenfishes, the largemouth or giant leatherskin (*S. commersonnianus*), grows to at least 14 kg (30 lb) in weight, with records of this size known from both northern Australia and South Africa. The largest queenfish listed by IGFA, a largemouth, was a huge specimen weighing 16.5 kg (36 lb) caught off Karachi, Pakistan, in 2002.

The other queenfish species do not reach these sorts of sizes, although the deep leatherskin has been recorded up to at least 7 kg (15 lb). On the other hand, the heaviest recorded double spotted queenfish weighed only 2.26 kg (5 lb) while the biggest needle-scaled queenfish weighed in at a rather unimpressive 0.5 kg (1 lb).

Virtually nothing is known about the basic biology of queenfish species. Some limited tagging in northern Australia has shown some restricted coastal movements, but unfortunately, no studies have yet been conducted on the reproduction of any of the queenfishes.

Queenfishes are commonly caught by coastal gillnet fisheries throughout Southeast Asia. Elsewhere, they are not regarded as important commercial species, but their value as a sportfish is well recognized. Needle-scaled queenfish are highly regarded as baits for marlin in some regions. Queenfish are particularly popular as target species of saltwater fly-fishing, and guiding operations through northern Australia and other areas take advantage of this growing attraction.

RAINBOW RUNNER
Elagatis bipinnulata

France	Comète saumon
Japan	Tsumuburi
Spain	Macarela

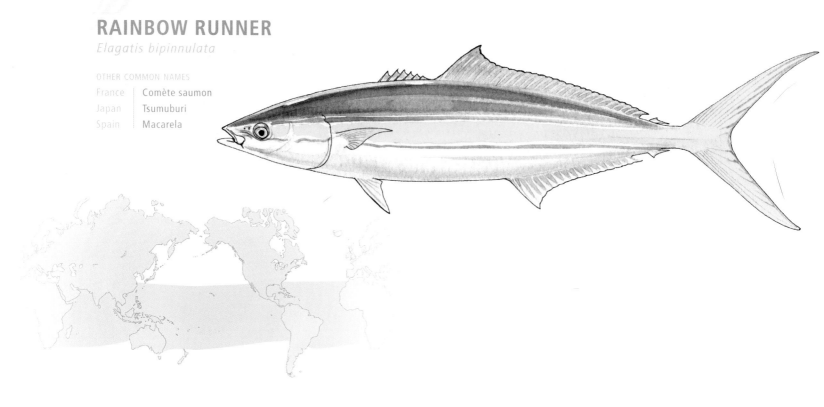

Like the dolphinfish, the rainbow runner, as its name suggests, is one of the very few open-ocean fishes that is highly colored. Its back is usually a deep purple, changing to flushes of royal blue as it 'lights up'. Two bright yellow, horizontal stripes run the length of the body, bordered by brilliant azure edges, presenting a stunning sight both in and out of the water.

A member of the trevally/jack family, Caringidae, the rainbow runner was once placed in the genus *Seriola* (the amberjacks), but is sufficiently different to be now allocated its own genus, *Elagatis*. The main difference between the rainbow runner and the *Seriola* species is that it has a pair of detached finlets (called pinnulae, hence its species name) on the tail wrist, whereas the *Seriola* species have continuous second dorsal and anal fins.

The rainbow runner occurs all over the world in tropical waters, although it is rare in the eastern Atlantic Ocean. While it is often found over reefs, it is also encountered far from land and is a true oceanic fish.

Very little is known about the biology of the rainbow runner. They sometimes aggregate with other species such as dolphinfish in association with floating logs and debris.

Rainbow runners do not grow very large, a 5 kg (11 lb) specimen being regarded as a good-sized fish. The IGFA all-tackle record was a 17.05 kg (37.6 lb) fish caught off Isla Clarion, Mexico. Another area where large specimens exceeding 10 kg (22 lb) are regularly caught is off Pinas Bay, Panama.

This beautiful tropical species is commonly encountered near reef dropoffs, and is a favored bait for marlin fishing. No directed fisheries target rainbow runner, although it is a common discarded bycatch species in fish aggregating device or log-associated purse seine fisheries. The quantities caught and discarded in these fisheries are unknown, but probably quite large. Even though it is commonly discarded, it is reported to be an excellent table fish.

Rainbow runners are very widespread and are common catches in artisanal fisheries in remote island areas. These were caught off Rabaul, Papua New Guinea.
Julian Pepperell

PILOTFISH
Naucrates ductor

OTHER COMMON NAMES
France | Poisson pilote
Japan | Burimodoki
Spain | Pez piloto

The pilotfish is a single worldwide species which occurs mainly in tropical to subtropical regions. It is a member of the trevally family, Carangidae, readily distinguished by its bluish body and striking black vertical bands, white tips on the tail and anal and second dorsal fins, and the possession of a fleshy lateral keel on the tail wrist (caudal peduncle).

Juvenile pilotfish are often found hovering with jellyfishes and drifting flotsam and jetsam, while adults are nearly always found associated with larger constantly swimming fish, especially sharks and large rays as well as marine turtles. Pilotfish take up a position in front of the leading edge of the head, dorsal or pectoral fins, apparently riding on the pressure wave. Juveniles of another carangid, the golden trevally (*Gnathanodon speciosus*), possibly mimic pilotfish, undertaking similar associations with large fish and turtles and being similarly marked with vertical black bands, albeit with a yellow rather than blue body.

Pilotfish grow to about 70 cm, but are most commonly observed at 30 cm or less.

9

OTHER PELAGIC GAMEFISHES

DOLPHINFISH

Coryphaena hippurus,
C. equiselis

OTHER COMMON NAMES

Cuba	Dorado
France	Coriphène
Hawaii	Mahi mahi
Japan	Toohyaku
Spain	Lampuga

Of all the fishes of the open ocean, the dolphinfish has been known to humans for the longest time. It is a relatively common inhabitant of the Mediterranean Sea, and the ancient Greeks and Romans were quite familiar with it, writing about its habits with obvious knowledge. Surprisingly accurate depictions of dolphinfish appear in Cretan frescoes and on Greek urns, so it is reasonable to assume that this fish was held in high regard. Mediterranean fishermen were well aware of the attraction of dolphinfish to floating objects, and used this to advantage, not only by fishing near floating debris, but also by constructing the first fish aggregating devices to make fishing for dolphinfish easy. As early as the mid-16th century, there is even evidence that aquaculture of dolphinfish was being attempted. The French naturalist Rondelet not only made accurate illustrations of the species, but gave accounts of Spanish fishermen keeping them alive in wicker pens for later sale at the markets.

The dolphinfish is one of the most colorful denizens of the surface layers of the world's oceans. Its vibrant hues flash from bright yellow to gold to silver to iridescent blue. When caught, they usually show yellow coloration but can also flash brilliant purples or even exhibit a mirror-like surface. There are two species of dolphinfish worldwide; the common dolphin (*Coryphaena hippurus*) and the somewhat rarer pompano dolphin (*Coryphaena equiselis*). The primary difference between the two is body depth – the common dolphin having its greatest body depth at about the level of the pectoral fin, while the pompano dolphin is deepest at mid-body. The pompano dolphin does not attain the size of the common dolphin. The two species are characterized by a single long dorsal fin, a deeply forked tail, broad pelvic fins and in the males, a very high forehead.

There is no connection between dolphinfish and the marine mammal dolphin.

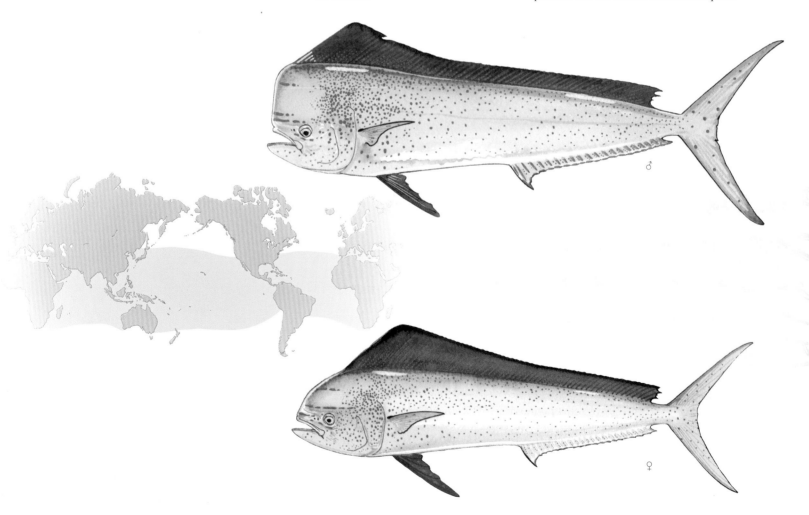

Of all pelagic
fishes, the
dolphinfish is
the most brightly
colored.
marinethemes.com/
Jen Kufs

The origin of the name dolphinfish is uncertain, but probably refers to its surface, fast-swimming habit, interspersed with arcing leaps, perhaps reminiscent of true dolphins.

Both species of dolphinfish occur in the three major oceans (Pacific, Atlantic and Indian), but the common dolphin has been recorded over the widest area. Dolphinfish are abundant throughout their range, especially around tropical and subtropical islands in all oceans.

Dolphinfish occur year-round in the tropics, but in more temperate latitudes their appearance is seasonal, coinciding with invasions of warm-water masses.

There are two main tagging programs under which large numbers of dolphinfish have been tagged. They are the Australian Gamefish Tagging Program, in which over 18000 common dolphinfish have been tagged off eastern Australia, and the program of the South Carolina Department of Natural Resources, in which 6000 dolphinfish have been tagged off the eastern United States in recent years.

In the Australian study, some 70 per cent of 174 recaptures were made at the points of release near fish aggregating devices or other buoys. The remaining recaptures indicated mainly short coastal movements, but some as far as 500 km. The exception to this rule, and the furthest distance moved between release and recapture for an Australian tagged dolphin was a remarkable 1810 nautical miles (3350 km) after 241 days at liberty. This fish was tagged north of Sydney at a size of only 40 cm and recaptured at Fiji weighing 9.8 kg (21.5 lb). This general lack of long-distance movements, especially offshore, is in direct contrast with results from the eastern United States. There, 156 tagged fish have been recaptured, and while many were also short-term recaptures, the average time at liberty was 40 days and average distance moved, an impressive 281 miles (520 km). One fish traveled 835 miles (1550 km) in only nine days, an average of 93 miles (172 km) per day, while another moved 137 miles (254 km) in a single day. And lastly, considering only tagged fish which moved away from the coast, the average distance traveled was a staggering 1560 miles (2890 km). These results are perhaps indicative of the level of fishing for dolphinfish throughout the Caribbean (and therefore the chances of recaptures) compared with the situation off eastern Australia.

A pompano dolphinfish (top) and a common dolphinfish caught in the same area off southeastern Australia. The pompano dolphinfish is deepest in the middle of the body and has a very short pectoral fin.
Tim Simpson

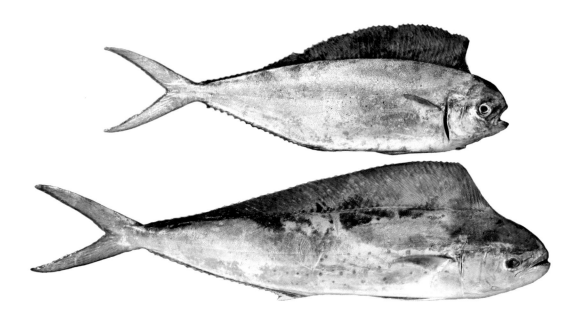

It is widely accepted that the dolphinfish is one of the fastest growing of all fishes. In captivity, male dolphinfish have grown from a pinhead-sized egg to over 16 kg (35 lb) in only eight months.

One classic study of the growth rate of the species was carried out off southern Florida, where scientists aged over 500 dolphinfish measuring between 47 cm and 1.52 metres. The results indicated that one-year-old fish ranged in size from 47 cm to 1.17 metres, two-year-olds (only nine fish) ranged from 1.0 to 1.13 metres, and that the largest fish, 1.52 metres long, was only three years old. At the time of the study, the all-tackle world record dolphinfish, weighing 35 kg (77 lb), was aged, and estimated at only four years old, the presumed maximum longevity of the species.

Female dolphinfish only grow to about half the maximum size of males. This is thought to be due to the constant production of eggs (see below), which consumes much of their available energy.

The all-tackle record for dolphinfish was a 39.46 kg (87 lb) fish, caught off Costa Rica in 1976, but a considerably larger specimen, weighing 46 kg (just over 100 lb), is reliably reported to have been caught off Puerto Rico in 1979.

Dolphinfish mature at a surprisingly small size. Off Florida, one study showed that female dolphinfish show signs of maturing at a length of only 35 cm, and are all fully mature at 55 cm or about 2 kg (4.5 lb). Similarly, the smallest maturing males average about 42 cm long. All fish of both sexes were found to mature within their first year of life.

Dolphinfish are relatively easy to keep in captivity and will spawn quite readily. Through these observations, it has been shown that dolphinfish will spawn every second day for many months on end, the females releasing perhaps 100 000 eggs at a time. Spawning takes place at the surface, nearly always at night, and another interesting observation is that fish usually form pairs to spawn.

Larvae of dolphinfish have been found throughout the tropics and sub-tropics of the world's three major oceans, and the Mediterranean. It would appear then that dolphinfish will spawn in any suitable location, given availability of food and surface temperatures above 23–24°C.

It has been known for thousands of years that dolphinfish have a strong attraction for floating objects. Why this is so is uncertain. It may be for protection, for feeding or for orientation in an otherwise featureless oceanic void, but whatever the reason, fishers have used this characteristic to their advantage, either by fishing near floating objects, or mooring buoys or floats especially designed for the purpose. This tendency to find and stay near floating objects is also a feature which makes dolphinfish very welcome fellow travelers of mariners around the globe, and of shipwrecked sailors who have survived by catching the dolphinfish attracted to their rafts.

The diet of dolphinfish is remarkably varied. A study of the items found in the stomachs of 2600 dolphinfish from the western Atlantic in 1984 found not only a wide variety of fish, crustaceans and squid in most stomachs, but also a significant quantity of plastic or other manufactured items such as tar balls, cigarette filters, nylon string and even a light bulb!

Dolphinfish are targeted commercially in the Caribbean, and many Pacific and Indian Ocean islands. The demand for dolphinfish in Hawaii (marketed as mahi mahi) is so high that large quantities need to be imported, with artisanal fisheries in the Philippines supplying much of this demand.

Sportfish anglers catch dolphinfish incidentally throughout much of their range. On the high seas, dolphinfish are a constant and significant bycatch of the longline fleets of the world. One of the problems with attempting to estimate the dolphinfish bycatch, especially of the distant-water longline fisheries, is that a high proportion of them are discarded at sea, simply because they would take up valuable freezer space which is intended for the main target species, bigeye and yellowfin tuna. This issue of unseen, unrecorded bycatch of important species such as dolphinfish certainly needs to be addressed by management agencies.

The dolphinfish is widely regarded as one of the best eating fishes which swims in the surface layers of the ocean. This is particularly true when eaten fresh, but the flesh tends to dry out somewhat when frozen. In some areas of its range, a protozoan parasite (*Kudoa* sp.) infects the flesh of dolphinfish, causing it to become soft and mushy after death, even when placed on ice immediately after landing. Fortunately, the phenomenon only affects a small proportion of fish. Because it occurs in the wild, however, the parasite has caused problems in the development of dolphinfish aquaculture industries in some areas.

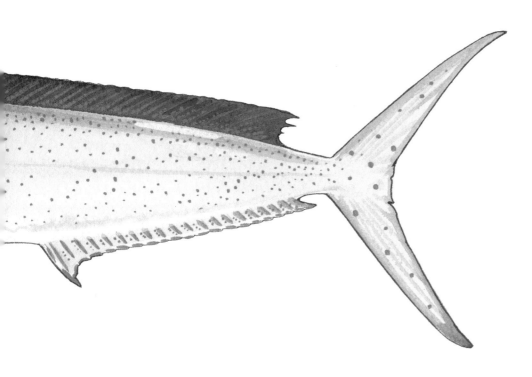

BARRACUDA

Sphyraena barracuda,
Sphyraena spp.

OTHER COMMON NAMES

	Great barracuda
Hawaii	Kupala
Japan	Oni-kamasu
Spain	Picuda barracuda

There are about 18 species of barracuda globally, although the taxonomy of the family (Sphyraenidae) is not fully resolved. Many of the species are relatively small, but several species grow to a large size, the largest of which is the great barracuda, *Sphyraena barracuda*. This is the most widespread of the barracudas, and the species responsible for much of the fearsome reputation of the group.

Careful anatomical studies of the barracudas have shown that they are part of the suborder of fishes known as Scombroidei, which includes all of the tunas, mackerels and billfishes. Furthermore, the barracudas have a number of so-called primitive features (such as mid-abdominal pelvic fins) which indicate that they quite probably resemble the ancestral fishes from which all of the more advanced tunas and tuna-like fishes evolved.

All of the barracudas are distinguished by a long, slender body which is nearly round in cross-section, a flattened head with a long snout and a markedly protruding lower jaw, very large, sharp triangular teeth and almost complete lack of gill rakers. The two dorsal fins are widely separated, and the paired pelvic fins are situated towards the middle of the abdomen. There are actually two rows of teeth in the jaws of barracuda – an outer row of small, razor-sharp teeth, and the much larger so-called 'canines'. These larger teeth fit into holes in the opposite jaw, such that the mouth can be completely closed – a formidable predator indeed.

The great barracuda usually has several scattered inky-black blotches on its rear flanks and belly, although this is not always the case. The tail of the great barracuda is broad and forked, with pale edges on both lobes. When seen underwater, the tail and second dorsal fin may appear jet black or violet, with white tips. The purpose of such striking coloration is not clear, although it would certainly divert the attention of a baitfish away from the head of the predator and its slashing teeth.

The great barracuda is a cosmopolitan (worldwide) species occurring in all three major oceans as well as the Mediterranean Sea. It is restricted to tropical/subtropical regions between about 30°N and 30°S. While its general distribution nearly extends across whole oceans (it appears to be absent from the eastern Pacific), in reality it is primarily concentrated near land masses, islands and

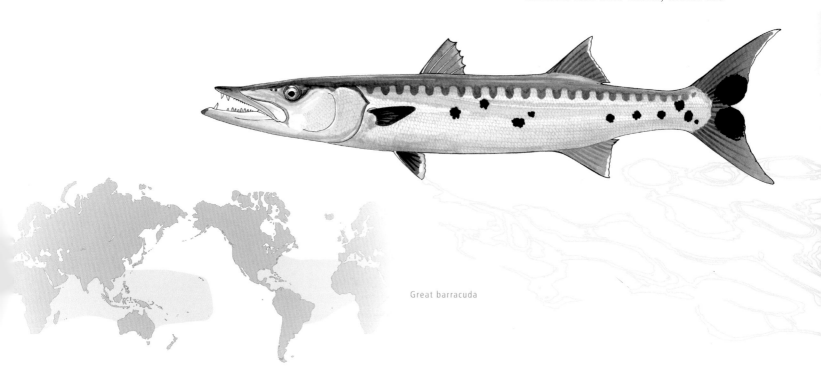

Great barracuda

The teeth of the great barracuda are as formidable as they look.
Julian Pepperell

atolls. As is the case with all of the barracudas, the early life stages of the great barracuda are spent in inshore sheltered habitats, especially among mangroves and seagrasses of tropical estuaries. As fish increase in size, they move progressively into deeper water, becoming pelagic and oceanic by full adult size, albeit with, as noted, a strong tendency to hunt near coral reefs and atolls.

The IGFA all-tackle record for barracuda is actually a tie between two great barracuda each weighing 38.5 kg (85 lb), one caught in Kiribati and one in the Philippines. However, this does not appear to be the species' maximum size. One reference to a fish of 44 kg (97 lb), measuring 1.7 metres in length, and caught off the Bahamas, appears to be quite valid, while a very early account of an even bigger fish is also verifiable. The fish was caught off Bimini, Florida, in 1932, and weighed 46.9 kg (103.5 lb). It measured 5 feet 6 inches (1.67 metres) in length and 30.5 inches (77 cm) in girth and was announced at the time to be the largest barracuda ever caught. It would appear that this was not only true in 1932, but is still the case.

Examination of the scales of great barracuda indicate that they reach about 60 cm long at two years of age, and may live for up to 14 years.

Reproduction of the great barracuda has mainly been studied around the Florida Keys, where it is believed that spawning occurs in offshore waters in the spring. Water temperatures above 23°C are required for maturation of gonads, each female producing many millions of eggs throughout the spawning season.

The barracuda is what is aptly known as a fast-start, or accelerator predator. This means that it can hold itself virtually motionless until it makes an explosive lunge towards an unsuspecting prey. This is achieved by the body shape of the barracuda – its broad, high tail with other fins set well back, a low, pointed head with long jaws, a wide lateral surface of the body and a high muscle mass to weight ratio. All of these features combine to provide a sudden, lethal forward thrust towards the target.

The great barracuda is esteemed as a gamefish in some parts of its range, and is actively pursued for sport and the table. This species, being a solitary hunter when adult, is not generally pursued commercially, although it is certainly a significant bycatch of many coastal gillnet fisheries in developing countries throughout its range.

Fear of barracuda

With the obvious exception of sharks, large barracuda undoubtedly instill greater fear in people than any other fish. Look up 'barracuda' in any encyclopedia and it is a fair bet that some mention will be made of their being dangerous to humans. This formidable reputation is based at least in part on fact, but the danger posed by the species has

Great barracuda
shadowing a
school of baitfish
among pylons.
Barracuda have
a broad habitat
range, from
shallow inshore
areas to open
ocean.
marinethemes.com/
David Fleetham

been blown out of all proportion and consequently has become something of an urban myth. No doubt, the appearance of this fish has a lot to do with its reputation. The large eye, and in particular, the long sharp wolf-like fangs, give the great barracuda a threatening look. Underwater, this is even more enhanced by the barracuda's habit of opening and closing its mouth, and by their ominous behavior around divers. Large barracuda have a habit of following closely behind divers, and while underwater attacks are rare, the experience of being followed can be unnerving. There are certainly attacks attributable to barracuda, but it is highly likely that these were accidental (from the points of view of both fish and humans). In some areas, attacks on humans by barracuda have been quite well documented. In the United States, for example, 19 confirmed barracuda attacks were recorded between 1873 and 1963. Two of these attacks apparently caused fatalities, one off Key West, Florida, in 1947, the other off North Carolina in 1957. The unfortunate Florida incident involved an airline pilot who was wearing fluorescent swimming trunks at the time. The bar-

racuda attacked his groin and buttocks repeatedly, and then hovered in the water nearby while the man, who later bled to death, was being rescued.

In Hawaii several incidents have also been recorded, including a case in Maui where a man casting a throw net was slashed on his leg by a six foot long barracuda. His injuries required five hours of surgery, while another man needed 255 stitches for wounds to an arm caused by a barracuda. Finally, two women in Kailua-Kona were bitten on the head by barracuda in separate incidents, one leaving teeth embedded in her scalp. Interestingly, both women were wearing shining head bands, a likely reason for the barracudas' interest. Shiny jewelry, including wristwatches, has been implicated in other attacks, so it is likely that flashes of light are at least partly to blame for the seeming aggressive behavior. It should be noted that there have been quite a number of incidents involving leaping barracuda, both hooked and free-jumping, causing injuries to people in boats. And it goes without saying that once a gnashing barracuda is brought into a boat, it pays to keep well and truly clear of the sharp end at all costs.

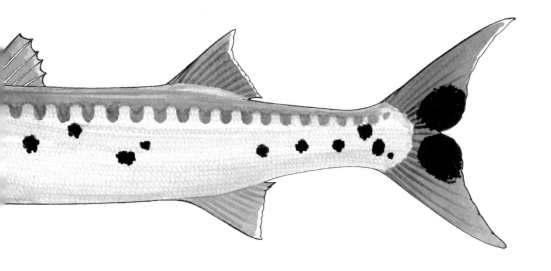

COBIA
Rachycentron canadum

OTHER COMMON NAMES

	Black kingfish
	Crab eater
France	Mafou
Japan	Sugi
Spain	Cobia

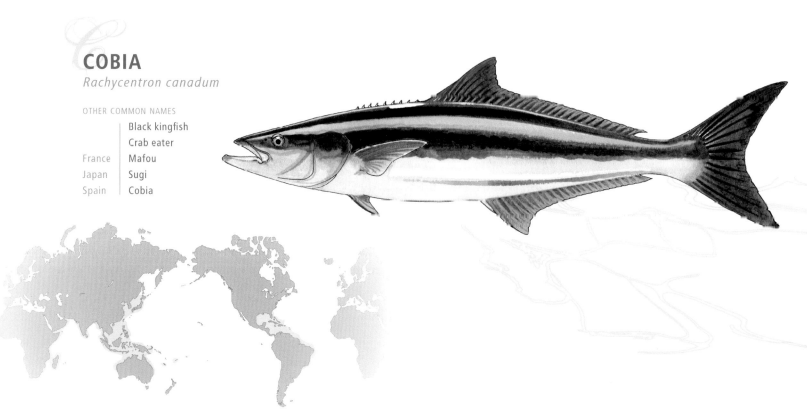

The cobia is the only member of the family Rachycentridae and has no other obvious close relatives. The early common name for this species in the Atlantic Ocean was actually 'cabio', so the name cobia would therefore appear to be a simple misspelling of the original name.

The body of this distinctive fish is cigar-shaped, with a long, flattened head and underslung lower jaw. Coloration varies between chocolate brown to charcoal, nearly always broken by two horizontal pearly or silvery-white stripes along each flank. The first dorsal fin consists of seven to nine very short spines which are not connected by a membrane – a distinct feature emphasized by its Latin name, Rachycentron, which means 'spine-backed'.

The cobia bears a superficial but striking resemblance to one other fish – the remora known as the slender suckerfish, *Echeneis naucrates*. The two are so similar in coloration and shape, especially when viewed from the side, that they could be easily mistaken for one another. There are two theories as to why this might be the case. The first is that the cobia, while not closely related to remoras, has evolved as a 'mimic' of the suckerfish, in a sense, fooling the larger fish with which they associate that they are in fact the remora to which the larger fish have become accustomed. The other theory is in effect the corollary of the first – that remoras evolved from the cobia (or cobia ancestor), eventually developing their sucking plates by the modification of the spiny dorsal fin. There is one problem with the latter theory, however. Studies of developing embryos and larvae of cobia and other fishes do not indicate that remoras and cobia are particularly closely related. Nevertheless, this mystery is still unresolved and awaits a definitive genetic study to examine the DNA of both remoras and cobia.

The cobia is one of those odd, single species of fish which is largely coastal in its habitat preference, but which is distributed virtually globally. Its preferred habitat is primarily confined to continental shelves of the major land masses, but it does extend its range in the Atlantic to islands in the Caribbean, and in the Indo-Pacific to larger islands such as Madagascar and Mauritius in the Indian Ocean and to New Caledonia and Fiji in the Pacific. Interestingly, the cobia is notably absent from the central and eastern Pacific, indicating that larger oceanic

distances must form an effective and permanent barrier to its movements.

Tagging of cobia has only been carried out in two areas – the US Atlantic/Gulf of Mexico coast and Australia. In the United States, tagged fish have been shown to move between the Gulf of Mexico and the southern Atlantic coast, and vice versa, and also from Florida to as far north as New Jersey. These movements suggest that the cobia off the eastern United States belong to one mixing stock. In Australia, of relatively few recaptures, the furthest distance moved by a tagged cobia is 145 nautical miles (270 km) along the eastern seaboard.

The most comprehensive study of the growth rates of cobia was undertaken in North Carolina in 1995. This indicated relatively fast growth, with males growing somewhat slower than females. Both sexes were found to reach a size of about 1 kg (2 lb) and a length of 50 to 55 cm in the first year, with males attaining about 74 cm by their third birthday and females, about 80 cm. The maximum ages of fish found in this study were 13 to 14 years, by which time males had

attained a weight of about 20 kg (44 lb), and females, 32 kg (70 lb).

The maximum size which cobia attain is not certain. The IGFA all-tackle record cobia weighed 61.5 kg (135.6 lb) and was caught at Shark Bay, Western Australia, in 1985. Other very large cobia which are listed as current records include fish of 58.4 kg (128.75 lb) caught off Florida in 1995, 56.9 kg (125.5 lb) off Florida in 1998 and 57.3 kg (126 lb) off Florida in 2000. Maximum sizes of 68 kg (150 lb) and 75 kg (165 lb) are variously quoted in the literature and, while such sizes may be possible, they are not able to be verified from the cited sources.

Cobia in the Gulf of Mexico have been shown to be batch spawners, with females producing between 370 000 and 1.9 million eggs every four days for several months. Spawning apparently occurs inshore in summer, near the mouths of large embayments on the Atlantic coast of the United States. Adult fish captured in these areas have spawned spontaneously shortly after capture and juveniles are sometimes found inside such embayments.

The cobia, the only member of its family, is an enigmatic species that occurs in a variety of habitats in all three major oceans.
Gary Bell/
OceanwideImages.com

TARPON
Megalops atlanticus,
M. cyprinoides

OTHER COMMON NAMES

	Silver king
France	Tarpon argente
Spain, Cuba	Sábalo

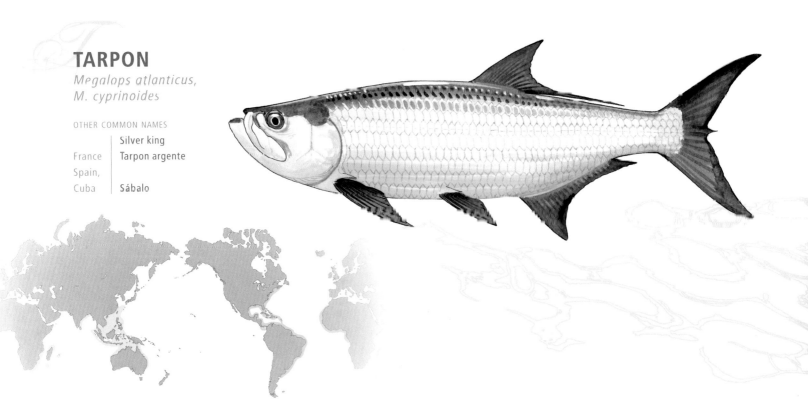

■ Tarpon
■ Oxeye herring

I dentification and classification of tarpon is simplified by the fact that there are only two species worldwide, one in the Atlantic and one in the Indo-Pacific. Both are in the family Megalopidae, the giant tarpon being *Megalops atlanticus*, while its smaller cousin, the oxeye herring, is *Megalops cyprinoides*.

The giant tarpon is found only in the Atlantic Ocean, ranging from Canada to Argentina in the west and from Senegal to Angola in the east. However, because it can tolerate fresh water, some stray tarpon have made it through the Panama Canal and a small population now exists on the Pacific coast of Panama. The smaller tarpon, the oxeye herring, is quite widespread, being found in South Africa, Zimbabwe, Madagascar, India, Southeast Asia, Papua New Guinea, northern Australia and some Pacific Islands, including New Caledonia, Palau and the Marianas.

Both species of tarpon can tolerate water with low dissolved oxygen by gulping air at the surface and storing it in their lung-like swim bladders. This allows them to penetrate into quite turbid water, including rivers and lakes some distance from the ocean, although, unlike the Atlantic tarpon, fully adult oxeye herring tend to avoid such habitats.

In common with quite a lot of fish species, female giant tarpon grow considerably larger than males. The IGFA all-tackle record for the species is 128.36 kg (283 lb), which is actually a tie – one fish taken in a lake in Venezuela in 1956 and the other more recently off Sierra Leone (Africa) in 1991. Two other specimens which appear in the line class records weighed over 120 kg (264 lb) and were also caught off Sierra Leone, obviously an area to target for large tarpon. The largest American line class record which is still in charts is a 66.8 kg (147 lb) fish caught on 4 kg (8 lb) line, although the saltwater fly world record is considerably larger at 91.85 kg (202.5 lb) for a fish caught off Florida in 2001. There is a reference in a Cuban paper to a tarpon measuring 2.5 metres and weighing 161 kg (354 lb). However, since this is so much larger than the all-tackle record, some doubt must be attached to the claim.

The Indo-Pacific species of tarpon, the oxeye herring, *Megalops cyprinoides*, is a much smaller fish. While sizes of 3 to 4 kg (7 to 9 lb) are not uncommon, and fish up to 8 kg (18 lb) have been caught in Australian waters, there are unconfirmed reports of the species reaching double this size – up to 18 kg (40 lb).

Giant tarpon grow quite rapidly

when young, reaching a length of about 1.2 metres in six to seven years, but slow down considerably after that. The age of a 100-pounder (45 kg), for example, is estimated to be about 15 years. Tarpon survive well in captivity, so we have some data on actual longevity of individual fish. Male fish are known to live to at least 30 years old, while one particular female fish, held in an aquarium in Chicago, died in 1998 at the ripe old age of 63. As with all cases of longevity in aquarium fishes, however, it is a moot point whether fish in the wild would attain such an age.

Being large fish, tarpon are highly fecund. It is estimated, for example, that a 2 metre female would produce approximately 12 million eggs at a time. Spawning peaks in summer months, although it is thought that some spawning occurs year-round.

Like the bonefish, tarpon larvae take the form of leptocephalic, ribbon-like creatures during their first several months of life. The whole cycle, from transparent leaf-like waifs, to juvenile fish measuring about 4 cm in length, takes approximately six months. During this extended period, it is very likely that larvae could be transported over large distances.

Tarpon do not reach sexual maturity until they are at least six years old – a factor which could contribute to their being prone to overexploitation.

Tarpon feed mainly on fish, including mullet, sardines, catfish and needlefish. They will also eat crustaceans, especially prawns and crabs. The wide gape of the tarpon's mouth creates a strong suction when it is suddenly opened, and when a large fish performs this noisy feat under a bait, it is a sight and sound not soon forgotten.

Some limited high-tech tagging of tarpon has been undertaken using popup satellite tags. Results have indicated that, in general, tarpon are more active during the day than at night and, interestingly, that peak activities occur immediately after cold fronts pass over the area. Other studies on the microchemistry of otoliths (ear bones) indicate that, while some tarpon migrate between the marine waters of the Caribbean and the fresh water of Lake Nicaragua in Central America, others do not, suggesting that movements are not predictable and may even be individually unique.

BONEFISH
Albula spp.

	Banana fish
	Lady fish
France	Banane de mer
Japan	Soto-iwashi
Spain	Macabi

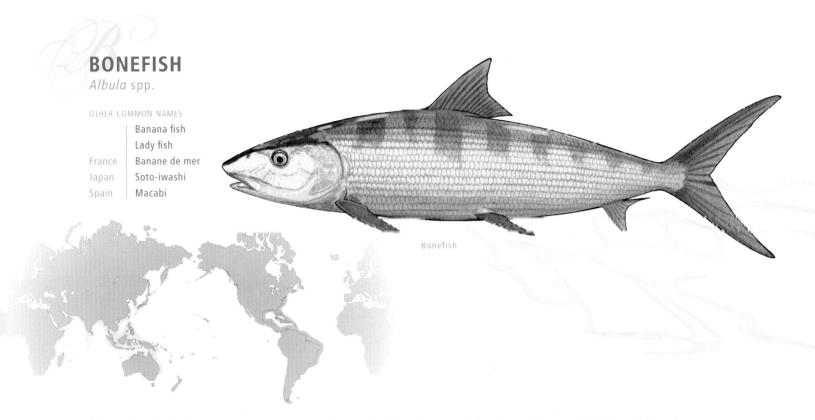

Bonefish

Roundjaw bonefish
Bonefish

Along with their distant relative the tarpon, bonefish are regarded as particularly 'primitive' in that they possess characteristics which are nearer to ancestral fishes than more recently evolved 'modern' fishes. These features include paired pelvic or ventral fins situated well back along the body, leptocephalic larvae (transparent ribbon-like creatures with very small heads) and a swim bladder which has a tube connecting it to the esophagus. The swim bladder is also highly vascularized and actually functions somewhat like a lung, enabling bonefish and tarpon to obtain extra oxygen by gulping air at the surface.

The primary species of bonefish which is prized by anglers the world over is *Albula vulpes*, literally (and aptly) translated as 'the white fox'. Until quite recently, it was thought that this worldwide species the Atlantic threadfin bonefish, *Albula nemoptera*, were the only bonefish species. However, recent studies have shown that there are as many as eight species of bonefish within the genus *Albula* and a further two in the genus *Pterothrissu*.

One species which overlaps with the 'true' bonefish is the roundjaw bonefish, *Albula glossodonta*. This species occurs through the tropical Indo-Pacific, and is

the most likely bonefish to be encountered in the western and central Pacific, including Hawaii. As the name suggests, the roundjaw has a distinctly rounded lower jaw, while that of the 'true' bonefish is more pointed. Another feature to note is the presence or absence of a small black spot under the snout. The roundjaw normally has this marking, while the 'true' bonefish does not. The threadfin bonefish, *Albula nemoptera*, which is confined to the Atlantic, is readily identified by its trailing rear extension of the dorsal fin (similar to that of the tarpon and giant herring).

Bonefish are cosmopolitan, meaning that they occur right round the world, mainly in the tropics. Their preferred habitat of shallow, sandy flats means that they are coastal when near continents, but they also occur widely, if patchily, in similar habitats around islands and atolls, often in remote mid-ocean locales. Well-known bonefish grounds include the Florida Keys, the Bahamas, Christmas Island in the mid-Pacific, and also in the eastern Indian Ocean, and the Seychelles in the northern Indian. In the western Pacific, New Caledonia, Fiji and Vanuatu also hold good populations of bonefish.

Growth rates of bonefish are not well

understood. Some estimates suggest that fish as small as 25 cm long may be as old as ten years, while others suggest that fish of this size may only be two years old. It is possible that these differences may be due to different species being studied.

Limited studies suggest that bonefish grow to different maximum sizes in different regions (again, this could be due to different species being studied). Off the southeastern United States and in the Caribbean, the largest bonefish reach about 77 cm in length and a weight of just over 7 kg (15 lb) – although any fish weighing more than 3.5 kg (8 lb) would be regarded as a prize specimen. The largest fish, however, have been recorded around oceanic islands and atolls. Fish weighing more than 9 kg (20 lb) are reported from Hawaii and western Africa, although just which species these might have been will most likely remain a mystery.

The IGFA world record bonefish was an 8.61 kg (19 lb) specimen caught off Zululand, South Africa, in 1962. Another weighing 7.7 kg (17 lb) was caught in the same region in 1976, while at least three fish weighing in excess of 7 kg (15 lb) have been taken around Florida. Two of these were caught on regular saltwater tackle, while the third, caught in 1997 on

4 kg (8 lb) tippet and weighing 7.03 kg (15.5 lb), is the current world fly-fishing record.

Bonefish spawn close to the coast, moving off the shallows into deeper water to do so. Off Florida and in the Caribbean, spawning takes place from about November to June, with little activity evident in the hotter months of the year. As mentioned above, bonefish possess a leptocephalus larval stage which reach a maximum length of about 60 mm – huge by larval standards. Metamorphosis to the more normal, and much smaller, larval form takes up to two weeks, and it is during this relatively long larval period that bonefish probably disperse throughout their broad range.

Most descriptions of the bonefish stress its 'inferior' (underslung) mouth which, together with the fact that the mouth is equipped with strong crushing teeth, indicate that this is a bottom feeder adapted to crushing shellfish. Indeed, dietary studies have shown that bonefish primarily eat crustaceans, including prawns and crabs as well as clams, mussels and oysters. Fish are also part of the diet, but these tend to be slow-moving burrowing species such as toadfish and frogfish.

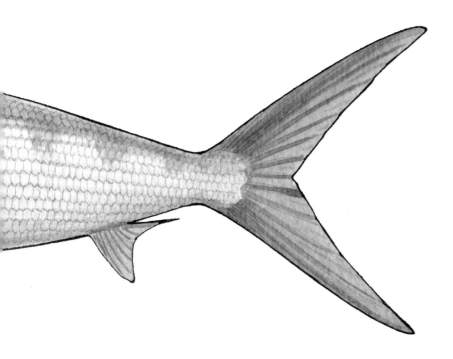

BLUEFISH
Pomatomus saltatrix

OTHER COMMON NAMES

	Tailor
	Elf
France	Tassergal
Japan	Okisuzuki
Spain	Anjova

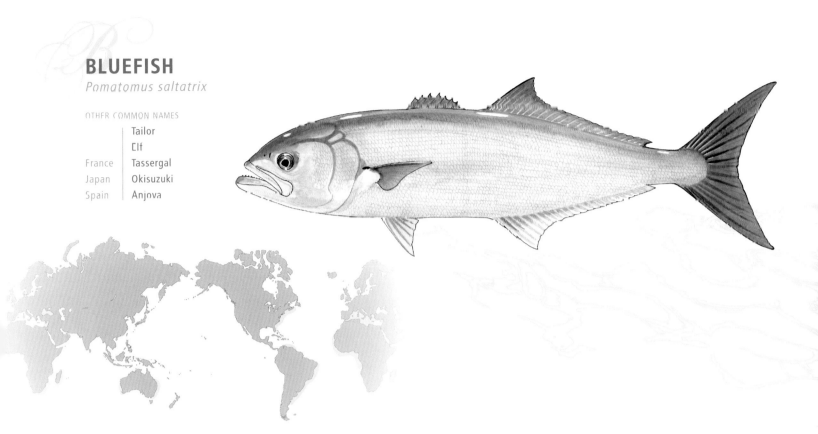

The bluefish is the only member of its own family, Pomatomidae. Its habitat is very much inshore – primarily along ocean beaches, rocky headlands and coastal islands. However, it is included here because of its highly unusual zoogeography and its interest to sportfish anglers around the world.

For a coastal species, the bluefish has a remarkable global distribution. It occurs right along the US Atlantic coast, and into the Gulf of Mexico, along the South American coast, from Venezuela to southern Brazil, off the southeast coast of South Africa, northwest Africa and into the Mediterranean, and along southeast and southwest Australia. The only temperate continental region it is absent from is the eastern Pacific. While mostly confined to coastlines of continents, bluefish also occur around some islands, notably the Azores and Madagascar. Global genetic studies show that the US Atlantic population is one stock, most closely related to the populations of Portugal and South Africa, while the Brazilian population was found to differ most from all other areas. Genetic differences between populations were found to be not great enough to warrant splitting one or more into different species, however.

A review of the biology of bluefish in different regions found a number of similarities. For example, adult fish undertake spawning migrations, after which eggs and larvae are carried by currents and juvenile fish enter inshore nursery areas at similar sizes. On the other hand, the number of spawning events, growth rates and reproductive parameters vary among populations. Bluefish grow fastest on the US Atlantic and northwest African coasts, and slowest in the Mediterranean and off South Africa and Australia. For example, studies have shown the length of fish at four years of age might range from 45 cm (Australia, South Africa) to 65 cm (northeastern United States). The maximum size attained is 1.3 metres and the maximum weight, as recorded by IGFA, was a fish of 14.4 kg (31.75 lb) caught at Hatteras, North Carolina, in 1972.

Being a coastal schooling fish with predictable migratory habits, the bluefish is easy to catch in large numbers and is therefore prone to overfishing. In response, management actions in different parts of the world have included seasonal and areal closures, including complete moratoriums on capture until stocks rebuild.

ROOSTERFISH
Nematistius pectoralis

OTHER COMMON NAMES

France	Plumière
Spain,	
Central America	Papagallo

For many years, the roosterfish was placed in the trevally family, Carangidae, which also includes the similar-looking yellowtail kingfish, *Seriola lalandi*, and amberjack, *Seriola dumerili*. However, close inspection of internal structures clearly showed that, while the trevallies and Seriolas were closely related, the roosterfish belonged in its own family, Nematistiidae, and further, that it is the only member of that family.

The obvious feature which sets the roosterfish apart from other fish, and which gives it its name, is the extraordinary dorsal fin, with its seven greatly elongated rays resembling an elaborate comb. The filaments of this fin fold neatly into a dorsal sheath, but when the roosterfish attacks its prey they stand out, perhaps to frighten or confuse the quarry. The markings of the roosterfish are also distinctive, showing two or three oblique dark bands against the silvery-white flanks.

Roosterfish are distributed along the Pacific coasts of Mexico, Central America and Peru, mainly in tropical waters, but extending to about 33° latitude north and south of the equator. While they are essentially a fish of the surf zone, they are sometimes encountered by offshore marlin fishers, and are also known to occur around the Galapagos Islands.

Tagging studies on roosterfish have been limited, but the results indicate little movement away from areas of release, the furthest recorded distance traveled being some 260 nautical miles (480 km) along the coast. Postlarval roosterfish about 1 cm long have been collected off Costa Rica in July, and juveniles have occasionally been found in rock pools and inside estuaries.

The all-tackle IGFA record roosterfish weighed 51.7 kg (114 lb) and was caught in 1960 off La Paz, Mexico. That fish was a giant, however, and any fish over 25 kg (55 lb) would be considered an outstanding specimen.

ATLANTIC SALMON

Salmo salar

OTHER COMMON NAMES

	Sea salmon
	Silver salmon
	Grilse
France	Saumon de l'Atlantique
Japan	Sake masu-rui
Spain	Salmón del Atlántico

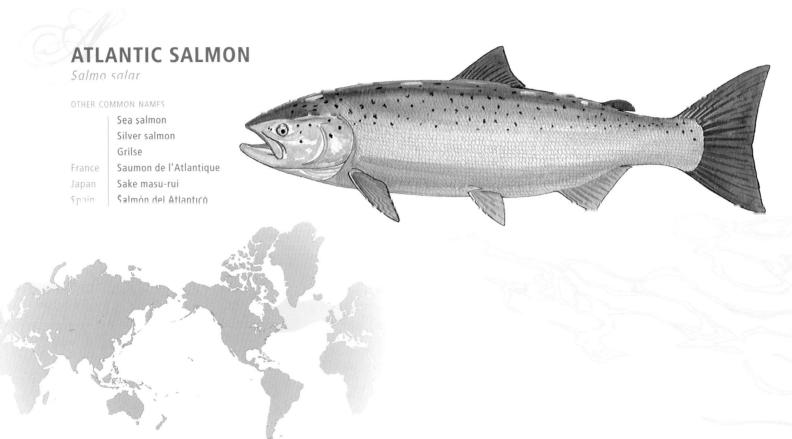

Two species of the family Salmonidae are included in this book because of the nature of their life cycles. Both the Atlantic and the chinook salmon begin their lives when they hatch from eggs deposited in freshwater rivers. They remain in fresh water for a number of years, but then undertake an age-old migration into oceanic waters, where they play important roles in cold-water open-ocean ecosystems. During this oceanic, pelagic phase, they feed actively, grow quickly and attain maturity before returning to their natal streams to begin the cycle anew.

The Atlantic salmon is relatively easy to identify in its native range since there are only a few salmonids which naturally occur in the Atlantic basin. The species has, of course, been widely introduced to freshwater systems around the world where it may overlap with many other salmonids, but here we are primarily concerned with the oceanic part of the life cycle of this species in its natural environment.

One feature which identifies the Atlantic salmon is the shape of the gill rakers on the first gill arch, all of which are sharp and thorn-shaped, whereas in the species most likely to be confused with the Atlantic salmon, the sea trout (*Salmo trutta trutta*), only the middle gill rakers have sharp points, the ones either side being blunt. The Atlantic salmon is also identified by its lack of spots below the lateral line. In marine waters, Atlantic salmon are silver, with a blue-green hue, darker on the back, and with a scattering of small dark spots over the dorsal surface but, as noted, not below the lateral line.

The native geographic range of the Atlantic salmon is confined to the far northern Atlantic, roughly between latitudes of 50°N and 70°N. In the west, this encompasses coastal rivers from the northern United States to Quebec, Canada, and in the east, from the British Isles through the Baltic States to Scandinavia and Russia. Western and eastern populations represent separate genetic stocks. The distribution in the open ocean is within the above latitudinal bands, from the surface to about 200 metres. The western shelf of Greenland appears to be a favored area where oceanic fisheries for the species take fish from both eastern and western stocks.

The Atlantic salmon has a fascinating life history that encompasses both the freshwater and marine habitats. As noted above, after growing slowly in fresh water for one to six years after hatching, most fish migrate to the open sea to feed and mature. This phase takes a further four or five years, after which fish make their way toward their natal river systems to spawn. Their movements while in the open ocean remain something of a mystery, however.

Atlantic salmon grow to an impressive size, with the largest recorded having been caught after they have matured and re-entered fresh water from the ocean. A maximum weight of 46.8 kg (103 lb) is given in the literature for this species, but further details regarding this figure proved elusive. The current IGFA record fish weighed 35.9 kg (79 lb), caught in fresh water in Norway in 1928. The fact that no larger fish has been claimed since then suggests that this must be near the upper limit for the size attained by the species.

A major difference between the reproductive biology of Atlantic and oceangoing Pacific salmons (including the chinook) is that, while the Pacific salmons invariably die after spawning once, only a small proportion of Atlantic salmon expire after spawning. The surviving majority return to the ocean again, to complete another spawning cycle.

When the urge to move to the open ocean occurs, fish swim strongly and purposefully to sea, as fast as two body lengths per second. Once in the ocean, Atlantic salmon spend much of their time at or near the surface feeding on a wide range of prey items, including crustaceans, fish and squid. In turn, they are heavily preyed upon by many predators, including seals, toothed whales and porbeagle sharks. It is thought that predation takes the greatest toll on salmon during their first few months at sea, perhaps because of the initially naive awareness of salmon to this new suite of predators. As a final comment, for such an important and iconic species, knowledge of its oceanic life phase is surprisingly poor.

CHINOOK SALMON
Oncorhynchus tshawytscha

OTHER COMMON NAMES

	King salmon
	Quinnat salmon
	Spring salmon
France	Saumon royal
Japan	Masunosuke
Spain	Salmón real

The second species of salmonid included here is the chinook salmon, one of a whole suite of Pacific salmon species which undertake migratory behavior from fresh water to the open ocean as an integral part of their life cycles. The chinook is widespread, grows to a large size (the largest of any salmonid) and is very important commercially and recreationally. It also has great cultural significance throughout its range through the coastal regions of the varied countries of the northern Pacific rim.

Identification of the Pacific salmonids can be rather daunting since the species resemble each other so closely. The chinook is most easily distinguished from others by the presence of scattered black spots on the back and on the tail, and by its black gums inside of the mouth (especially of oceanic fish). Oceanic fish also have silvery sides with a blue-black back.

The native geographic distribution of the chinook salmon is across the northern Pacific, mainly between latitudes 45°N and 65°N but extending as far south as northern California in the east. The species extends along the southern coast of Alaska and the Aleutian Islands through to Russia and Japan. During their oceanic life phase, chinook salmon spend much of their time at or near the surface, but may be found on occasion in depths of up to 375 metres.

Chinook salmon migrate to the sea quite early in life, sometimes within just months after hatching, but usually between a year and eighteen months. Some individuals, however, may stay in fresh water for up to three years. Not all fish undertake this migration (those that don't are called 'jacks'), but most do, expanding into broad oceanic areas of the far northern Pacific. During the oceanic life phase, fish feed and reach maturity. This may take from three to five years, after which they unerringly return to their natal stream to spawn, and afterwards, to die.

As mentioned, this impressive fish is the largest of the salmonids, growing to a length of about 150 cm and a reported weight of 61.4 kg (135 lb). The IGFA record is considerably lower than that figure, being a 44.11 kg (97 lb) fish caught in Alaska in 1985. As with many of the salmonids, it is very difficult to be definitive about growth rates since they depend so much on many factors. These include

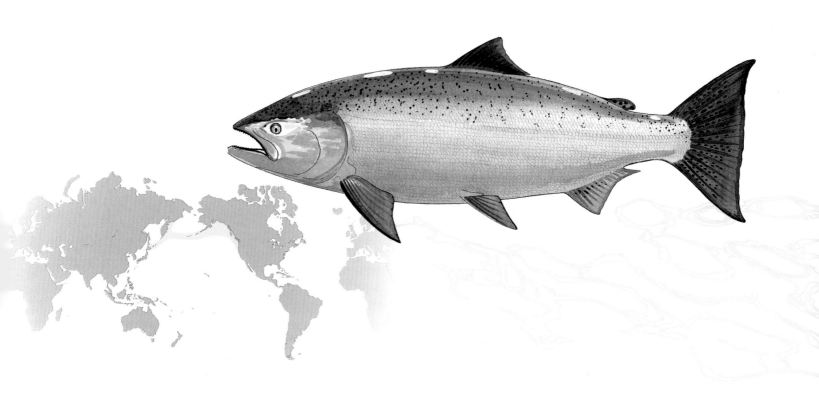

environmental conditions, availability of food and the genetics of the stock. Having noted these variables, the average life span of a Chinook salmon would be four to five years, extending to nine or ten years at the extreme.

After growing and maturing at sea, chinook salmon return to the coast as adults averaging perhaps 12 to 14 kg (25 to 30 lb) in size, where they enter their natal catchment and make their way up the river to spawn. These spawning runs can be long, tortuous journeys, some of which may be more than 3000 km long. Once spawning has taken place, as is the case for all of the Pacific sea run salmons, Chinook salmon undergo 100 per cent mortality.

Surprisingly little is known about the oceanic part of the life cycle of chinook salmon. The importance of the species in the ecosystem of the northern Pacific is unclear, although the biomass of chinooks (and other species of Pacific salmon) which reside in the northern Pacific at any one time must be substantial.

Like the Atlantic salmon, chinook salmon eat a wide variety of prey organisms, including a preponderance of crustaceans. Certain chemicals within crustacean exoskeletons are absorbed by the salmon, causing the flesh to take on the characteristic orange-pink color. Oceanic chinook salmon would be preyed upon by many predators during this phase of their lives. These would include many marine mammals but would also include in particular the salmon shark, *Lamna ditropis*, the life cycle of which is intimately linked with the salmon. This is shown not only by the presence of various species of Pacific salmon in the stomachs of salmon sharks, but by the fact that the abundance of the sharks in the Aleutian Islands and the Gulf of Alaska corresponds with abundance of salmon, as indicated by commercial catch rates.

After spending several years in the open ocean, these Alaskan chinook salmon make their way unerringly to their natal stream where they will spawn and die, once again completing, and beginning, an age-old cycle.
boyceimage.com

10

SHARKS AND RAYS

WHITE SHARK
Carcharodon carcharias

OTHER COMMON NAMES

	Great white shark
	White pointer
	White death
France	Grand requin blanc
Japan	Hohojirozame
Spain	Jaquetón blanco

Of all the potentially dangerous creatures in the ocean, the white shark is probably feared the most. It is the largest predatory fish on the planet, and when adult, its primary prey items are marine mammals. It is known to have been responsible for many attacks on humans, a relatively high proportion of which have been fatal. Early European records of white sharks are somewhat dubious, although there is little doubt that the species was well known as a dangerous shark from ancient times.

The white shark is a member of the Lamnidae family (mackerel sharks), and shares with this group a very robust body, broad caudal keel, large gill slits extending to the dorsal surface and a nearly symmetrical crescent-shaped tail. It may be distinguished from the other lamnid sharks (makos, porbeagle and salmon shark) by its teeth, which are triangular and serrated on both margins.

The white shark is a true cosmopolitan (worldwide) species, with a geographic distribution covering an enormous area. It is found along the coasts of the main land masses of all three major oceans, in the Mediterranean and Red Seas, and around many offshore islands, including New Zealand, Hawaii, the Marshall Islands, New Caledonia, the Azores and the Maldives. Its range extends as far north as the top of the Gulf of Alaska, and as far south as southern Argentina. While white sharks are generally considered to be creatures of cooler waters with preferred temperatures below 18°C, large adults certainly penetrate into the tropics. Newly born pups and juveniles, on the other hand, are nearly always found in temperate waters.

Until recently, white sharks were not considered to be highly mobile, with tag returns and sonic tracking studies suggesting that they stay near the coast in relatively shallow water over continental shelves. Conventional tagging has been undertaken off South Australia and South Africa, and in both cases, most movements recorded have been within a few hundred kilometres of release points. However, recent studies have completely altered this view. A juvenile (2 metre) white shark tagged off the coast of Victoria, Australia, with a satellite transmitter attached to its dorsal fin undertook a remarkable journey over the ensuing three months. It traveled across Bass Strait to the eastern coast of Tasmania, before returning to the mainland, and traveling north along the southeastern mainland coast for at least 1000 km.

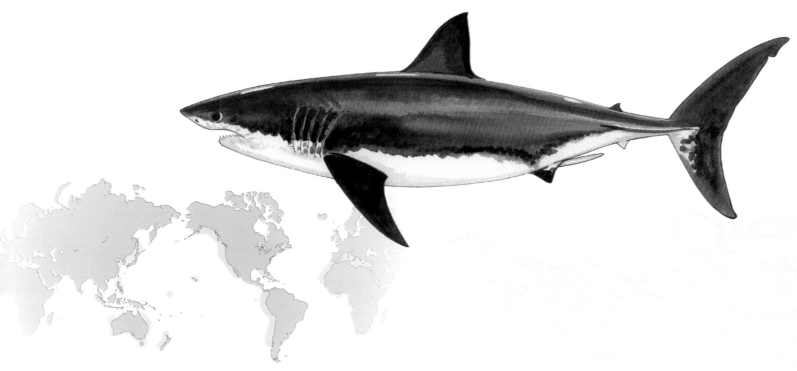

The second, even more dramatic study involved the tagging of six adult white sharks off the coast of California with popup satellite tags designed to record a range of parameters such as daily location, depth of diving and water temperature, and then to release from the fish after a predetermined length of time and transmit the stored data to a satellite. The results showed that four of the sharks left the coast and headed out into the eastern Pacific. One traveled at an average speed of 70 km per day for over 5000 km to reach Hawaii, where it stayed for four months. The other three sharks all moved to an area in the subtropical Pacific more than 2000 km west of Baja California, where they also remained for several months. During these movements, the latter sharks remained near the surface, but exhibited intermittent deep dives. The purpose of these seemingly purposeful horizontal movements, and deep dives is thought to be related to feeding. In fact, the area in the Pacific to which the three white sharks headed has been dubbed the 'shark café', since two other lamnid species, shortfin makos and salmon sharks, have also been shown to travel great distances to reach this same location.

More recently, a white shark that was satellite tagged off South Africa crossed the Indian Ocean to the western coast of Australia, and then returned to South Africa, while another tagged off eastern Australia crossed the Tasman Sea, swam between the two main islands of New Zealand and reached the Chatham Islands, 800 km further east.

There have been a number of attempts to estimate growth rates of white sharks. A South African study with the largest sample size – 113 fish ranging from 42 kg (92 lb) to 882 kg (1940 lb) – estimated that growth rate was moderate. The smallest shark in this study was almost certainly one year old, while the largest, a male, was estimated to be 13 years old.

The maximum size to which white sharks grow is a topic which has aroused considerable debate. There are numerous tales of huge white sharks measuring over 30 feet in length being caught, and there are even photos and jaws of such apparent monsters. However, when such stories, and the 'hard' evidence have been examined by scientists, the actual sizes of the sharks shrink considerably. After considering all of the available evidence, two respected scientists, Dr Jack Randall and John McCosker, both came to the conclusion that the maximum length of any white shark which was actually measured was probably between 21 and 23 feet (6.4 to 7.0 metres). The largest white shark which has been completely verified, was one caught in a noose off Western Australia in 1984. It measured 19 feet 6 inches (5.94 metres) and weighed 1511 kg (3324 lb). The heaviest officially measured weight, however, is greater – 1557 kg (3427 lb) for one caught off Montauk, USA, in 1986, although, interestingly, that shark only measured 17 feet (5.2 metres) in length.

The reproduction of white sharks has been difficult to study due to the scarcity of observations of pregnant females and hence, their scientific examination. Counts of embryos in the few white sharks examined have ranged from two to 14, although the counts at the lower end of this range are thought to be possibly due to spontaneous abortion of most of the pups during the act of capture. Detailed inspection of a 5.15 metre pregnant white shark caught in Japan revealed eight full-term embryos ranging from 135 to 151 cm long and from 21.3 kg (47 lb) to 32.4 kg (71 lb) in weight. The stomachs of the embryos were filled with yolk, clearly indicating that they had been feeding on eggs within the uterus – a strategy termed oophagy which is also known in the related mako and porbeagle sharks, as well as in the unrelated thresher and ragged tooth sharks.

Size at first maturity is probably about 3.5 metres for males, and perhaps 4.5 metres for females. Corresponding ages would be about eight to nine years old for males and 12 to 13 years old for females.

White sharks change their food preferences with size. When small, they tend to feed primarily on fish, but on reaching

The unmistakable
profile of the
white shark,
perhaps the
ultimate oceanic
predator
boyceimage.com

an adult size of 3.5 to 4 metres long, the diet largely shifts to marine mammals, especially seals. Large white sharks also prey on dolphins and are known to attack newborn southern right and humpback whales, and to feed voraciously on floating carcasses of whales of a variety of species.

Apart from demonstrating unexpectedly long oceanic movements, the satellite tagging studies referred to above also showed that white sharks dive to much greater depths than previously thought. The maximum depth recorded by one of these electronically tagged sharks was 680 metres below the surface, while most animals spent significant amounts of time between 300 and 500 metres. The other preferred depth stratum was within 5 metres of the surface, and all four oceanic traveling sharks in fact spent 90 per cent of their time at those two depth strata. The temperature range between these depths was very broad, ranging from 26°C at the surface to 4.8°C at the greatest depth.

White sharks have now been protected in South Africa, Australia, California and the Maldives (although few are found in the latter area). Nevertheless, hooking, netting or entanglement in commercial fishing gear in various parts of their range is thought to still cause substantial mortality. Protective beach meshing and baited hook programs in South Africa and Australia also cause some mortality of white sharks, although the total global human-induced mortality rate is difficult to determine. In some areas, it is thought that white shark numbers are on the increase, especially since the abundance of seals and whales has certainly increased over the past several decades.

MAKO SHARK

Isurus oxyrhinchus,
I. paucus

OTHER COMMON NAMES

	Blue pointer
	Mackerel shark
	Bonito shark
France	Taupe bleue
Japan	Aozame
Spain	Marrajo

Longfin mako
Spain | Marrajo carite

The mako sharks belong to the mackerel shark family, Lamnidae, the members of which (white, porbeagle, salmon and mako sharks) are widely regarded as the most 'advanced' of all 400 plus species of sharks worldwide. Apart from their similar peak 'design' in body form, they all share a very unusual feature: they are to some extent warm-blooded. In fact, their internal biology in some ways is closer to that of mammals than fish, proving that these sharks have certainly evolved to a high degree, even though their external body shape may have changed little over the ages.

There are two species of mako shark, by far the most common being the shortfin mako, *Isurus oxyrhinchus*. As the name suggests, the longfin mako, *Isurus paucus*, has very long pectoral (side) fins which are at least as long as the head (measured from the tip of the snout to the last gill slit), whereas the pectorals of the shortfin mako are shorter than the head length. The longfin has a somewhat blunter snout than the shortfin, and also has a larger eye, although in both of these cases, these features are not obvious without a shortfin alongside for comparison.

While makos are distinctive sharks, they can still be confused with the three other lamnid sharks – the porbeagle, salmon shark and white shark. The primary feature which separates the white shark from the mako is the teeth. Whites

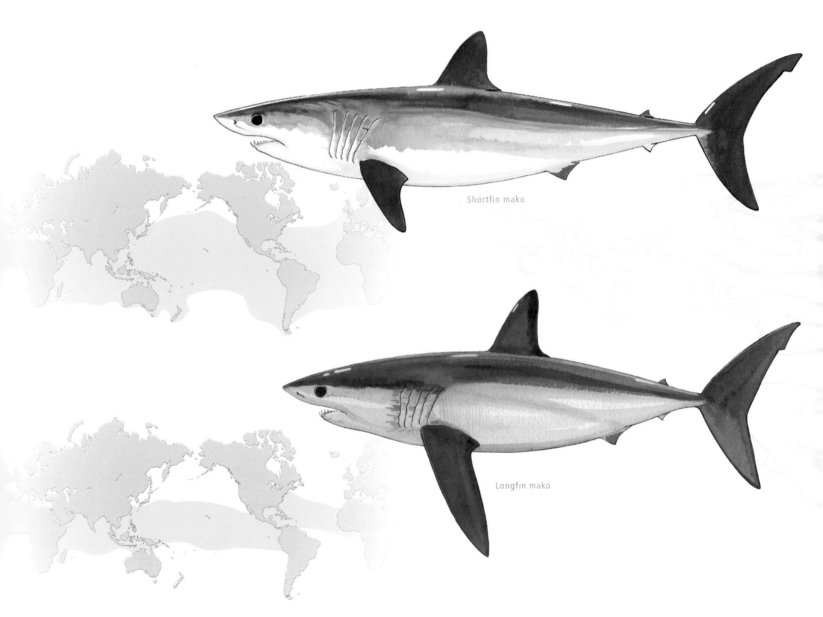

Shortfin mako

Longfin mako

have triangular, finely serrated teeth in both jaws, whereas the mako's teeth are dagger-shaped, and curve inwards. Porbeagle and salmon shark teeth are not serrated, but have two cusps at the base of each, which the mako lacks. There is a white patch on the rear free trailing edge of the porbeagle shark, while the mako's is uniformly colored. Lastly, the salmon shark can be separated from the makos by a whitish patch which extends from the belly to above the base of the pectoral fin.

The extended pectoral fins of the longfin mako suggest an oceanic, current-gliding existence, much like the blue shark, while the large eye indicates that the longfin is probably a deep-water hunter. Apart from these speculations, virtually nothing is known about the biology of the longfin, so this treatment naturally concentrates on the shortfin mako, which has been the subject of a number of studies.

The shortfin mako has a truly worldwide distribution, extending into temperate waters in all three major oceans. The lower limit of temperature tolerance for makos is usually quoted as 16°C, although off the northeastern United States, large ones are sometimes encountered in waters as cool as 11°C.

Makos are generally described as tropical to temperate sharks, which is a little confusing to anglers since they are rarely caught in surface waters in the tropics. The reason for this description is that Japanese longline fleets historically captured makos throughout the tropics, and mapped this distribution. However, longline hooks fish well below the surface, where temperatures are considerably lower than at the surface. Like many other pelagic species which prefer cooler waters, mako sharks simply spend most of their time below the thermocline in the tropics to avoid the hot, surface layer.

Large numbers of mako sharks have been tagged by recreational anglers in several locations around the world, resulting in some very interesting information on their movements. The main tagging has occurred off New Zealand, the eastern United States and southeastern Australia. From such tagging in each area, some

lengthy movements of makos have been recorded, but the overall results suggest some site fidelity and structuring of stocks within both the Atlantic and the Pacific oceans.

By far the greatest number of mako sharks have been tagged off New Zealand. A total of 11 300 makos had been tagged by the end of the 2006 season and 316 recaptures had been reported. The furthest distance moved to date by a New Zealand tagged mako has been to the Marquesas Islands, near Tahiti, a distance of 3000 nautical miles (5500 km) after 162 days of liberty. Eleven makos tagged in New Zealand have been recaptured in Australian waters. Two of these had been at liberty for over two years, but one had made the trans-Tasman crossing in only 36 days, at a minimum rate of travel of 27 nautical miles (50 km) per day. An interesting finding from the New Zealand program is that around 15 per cent of all recaptured makos have been caught well to the north in Fijian waters, nearly all by longline vessels. This cluster of recaptures seems to indicate a seasonal movement of makos into warmer, tropical waters during the southern winter, although other recaptures show that not all makos undertake this journey.

The Australian Gamefish Tagging Program has seen over 6000 makos tagged, virtually all off New South Wales, and a total of 142 recaptures have been recorded. Movements of makos tagged on that program have been predominantly coastal, but some long-distance displacements have been recorded. Makos tagged off southeastern Australia have been recaptured in the Coral Sea, while international movements have been recorded to New Zealand and some Pacific Islands. The record distance moved so far was 2577 nautical miles (4755 km) by a mako tagged off southeastern Australia and recaptured in the Philippines 585 days later.

Until the mid-1990s, about 3500 makos had been tagged off the eastern United States and 320 recaptures reported – a high recapture rate of 9.3 per cent. Makos tagged off the eastern United States have been recaptured in the central

Mako sharks are
just as active at
night as during
daylight hours.
boyceimage.com

Atlantic, the Gulf of Mexico and several as far south as Venezuela, but all to the north of the equator.

Several features of the movements of makos are evident from all three tagging programs. Significantly, there has been very little movement of fish between northern and southern hemispheres in either the Atlantic or Pacific oceans, and there has been very little movement recorded from the western Pacific or from the western Atlantic into the eastern sides of either ocean. This is so marked in the Atlantic (a much smaller basin than the Pacific) that it is theorized that the mid-Atlantic ridge must form some kind of barrier to the movement of makos across that ocean.

These results all suggest that the populations of mako sharks in both the Pacific and the Atlantic form discrete stocks, both north and south of the equator, and east and west in both oceans. Genetic studies on makos have been conducted, but have not been able to detect any differences in populations, even on a worldwide basis, but in this case, tag results strongly support the idea of quite separate populations, even though there may be some genetic 'leakage' over long periods of time.

The growth rate of mako sharks has been estimated in a number of studies, mostly by counting concentric rings visible in vertebrae. In the eastern United States, detailed examination of vertebrae collected throughout the year has shown that it is likely that two rings are laid down each year. However, a more recent study from South Africa, based on a single shark which had absorbed radio isotopes when very small, suggested that only one ring is laid down annually. If two rings per year are laid down, then studies indicate that makos grow quickly in their first few years, from about 70 cm (4 kg – 9 lb) at birth to about 180 cm (60 to 80 kg – 130 to 175 lb) by the end of their third year. As well, big makos would not be particularly old – a 320 kg (705 lb) female being estimated at about ten years old. Of course, if one ring is laid down each year, then these estimated ages would be doubled.

The shortfin mako shark with its characteristic pointed snout and large black eye. Note also the recurved dagger-like teeth, suited to gripping rather than slicing. *marinethemes.com/ Richard Herrman*

As with many species of shark, the maximum size to which makos grow is somewhat evasive. The all-tackle world record shortfin mako weighed 553.8 kg (1221 lb) and was caught off Massachusetts, USA, in 2001. This surpassed the previous record of 505 kg (1115 lb) taken off the island of Mauritius in the Indian Ocean in 1995. Two other thousand-pounders appear in the IGFA charts; one weighing 490 kg (1080 lb) caught off New York in 1979, and the other weighing 488 kg (1075 lb) caught off Spain in 1997. The largest recorded Australian mako, caught off New South Wales in 1991, weighed 467 kg (1027 lb), while in New Zealand, the apparent 'home' of the mako, the all-tackle record which has stood since 1970 weighed 481.26 kg (1059 lb). These are official records, but there is little doubt that there have been larger makos caught and verified. The largest appears to be one taken off Massachusetts in July 1999 which weighed 602 kg (1324 lb). This particular fish was caught on rod and reel but not claimed as a record for technical reasons. The largest longfin mako measured 4.17 metres. It was caught off Florida in 1984.

The mako sharks, like the other lamnids (mackerel sharks), are ovoviparous, meaning that the eggs hatch inside the mother well before birth. The developing young then eat other eggs as they are produced by their mother, their abdomens becoming incredibly distended as a result. (This behavior is technically known as oophagy, or 'egg eating'). This reproductive strategy sounds quite bizarre, but biologically, it makes perfect sense. It means that the young which win the race within the uterus can grow to a much larger size before birth than would be the case if each embryo developed from the yolk of one egg. This is reflected in the size at which baby shortfin makos are born – about 4 kg (9 lb), much larger than the newly born tiger or blue sharks which are not oophagous.

Pregnant mako sharks have not been scientifically examined very often. For a long time, this meant that there was considerable uncertainty about the size at first maturity of makos, where they pupped, the numbers of pups they produced, and the size of the pups at birth. Fortunately, a study in the United States collected data from around the world on any reliable observations of these sorts of variables and summarized all of the data. For 30 pregnant mako sharks recorded, the number of pups found ranged between four and 28, with an average of 12.5. (The authors of the study make the point that some of the counts of near full term embryonic makos are thought to be on the low side because of the possibility of a portion of the litter being aborted during capture.)

The average number of pups, at 12.5, is somewhat higher than the shortfin mako's close relatives. The longfin mako apparently gives birth to an average of four young, great whites, 8.9 and the porbeagle, four. These numbers are based on very small sample sizes, but they do give an indication that the shortfin mako is the most fecund of the lamnid sharks.

The size (fork length) at first maturity of female makos has been estimated at between 2.7 and 2.9 metres, which, based on the aging studies mentioned above, would equate to an age of six to eight years (for two vertebral rings per year) or 12 to 16 years (for one vertebral ring per year).

Mako sharks are very good to eat. This fact has meant that they are targeted in earnest by commercial fisheries in some parts of the world. It is not unusual to see mako on the menu in American seafood restaurants, and Spanish longline fleets in the eastern Atlantic specifically target mako and thresher sharks, both of which are caught in large numbers. Throughout the Pacific, the longline fleets of many countries take unknown quantities of mako (including longfin). These catches are often not recorded in any way, but it is likely that the total catch is large.

PORBEAGLE SHARK

Lamna nasus

OTHER COMMON NAMES

	Mackerel shark
	Blue dog
	Beaumaris shark
France	Requin-taupe commun
Japan	Mokazame
Spain	Marrajo sardinero

The porbeagle shark is a member of the lamnid or mackerel shark family, which also includes the mako, white and salmon sharks. All of these sharks have been shown to have some thermoregulatory ability, by means of which internal body temperatures are maintained above those of the surrounding water. Porbeagles in particular are thought to have developed this capacity to a high degree, explaining how they have been able to penetrate the colder regions of the three major oceans.

The porbeagle is generally thought of as a northern Atlantic species, although it also occurs circumglobally in subantarctic and temperate waters of the southern hemisphere, including the southern coast of Australia and around New Zealand.

Porbeagle sharks are very robust, stocky sharks with a markedly pointed snout and a large, almost symmetrical tail. They can be distinguished from the other sympatric lamnid sharks, makos and white sharks, by the following diagnostic (unique) features. The free trailing edge at the rear of the dorsal fin of the porbeagle has a pale to white patch, whereas the dorsal fins of whites and makos are uniform in color. This is an excellent feature to look for if a porbeagle is suspected, and can be seen clearly even without removing the shark from the water. Porbeagle teeth are similar shapes in both jaws. Each tooth is quite slender, smooth-edged, and importantly, has a small cusp on either side at the base, whereas whites and makos lack this cusp. Lastly, the porbeagle has a small, secondary keel below the main lateral keel on the tail wrist (caudal peduncle), while white and mako sharks possess only the main keel.

The maximum size recorded for a porbeagle is 350 cm total length, and the maximum weight, 230 kg (507 lb). Porbeagles have been aged by counting concentric rings on vertebrae, which correlate with size frequency analysis. A study which validated the frequency of formation of growth rings (a pair of rings is laid down each year) showed that porbeagles have a very slow rate of growth. Born at a length of 70 to 80 cm, they take ten to 12 years to reach a length of 200 cm. All evidence points to this being a very long-lived shark, with longevity estimates ranging from a conservative 30 years to perhaps 65 years.

Porbeagles are ovoviviparous with litter sizes ranging from one to five (usually four). The embryos are oophagous, but it is uncertain whether they are intrauterine 'cannibals'; in other words, if they eat their developing brothers and sisters or not. Certainly, they gorge themselves on fertilized eggs which the mother produces throughout gestation, midterm embryos having enormously distended stomachs as a result. Gestation is thought to be about eight months. Pregnant females have been found in all months off eastern Canada and off Europe from October through June, suggesting that there is no extended resting period between pregnancies (which is the case for some other shark species). Males reach maturity at about eight years of age and females at about 13 years. The growth rates of both sexes slow down after reaching maturity, but because males mature earlier than females, the largest mature individuals are always female.

Together with bluefin tuna and their sister species, the salmon shark, porbeagles are the ultimate predators in the subtemperate pelagic waters of the world.

Their preferred temperature is between 5°C and 10°C, but because they are able to thermoregulate by means of a countercurrent heat exchanging circulatory system, their core body temperature is maintained up to 16°C above that of surrounding water. Being warm-blooded to this extent is thought to convey an advantage of porbeagles over their cold-blooded prey species. Porbeagles prey on a wide range of organisms, including blue mackerel, herrings, cod, dories, dogfish and squid. No attacks on humans are attributable to porbeagles, although there may have been some incidents of porbeagles biting boats. The species is regarded as potentially dangerous.

The flesh of the porbeagle is particularly good to eat, and consequently, they have been a target species of commercial fleets in the north Atlantic. There have been concerns about the status of their stocks for many years, and catches are now only fractions of earlier harvests. This vulnerability to overfishing is well known in many shark species, particularly those like the porbeagle which have a very low rate of generational turnover.

The porbeagle is a colder-water shark that occurs in both the northern and southern hemispheres.
Andy Murch/
imagequestmarine.com

SALMON SHARK
Lamna ditropis

OTHER COMMON NAMES

France	Requin-taupe saumon
Japan	Nezumizame
Spain	Marrajo salmón

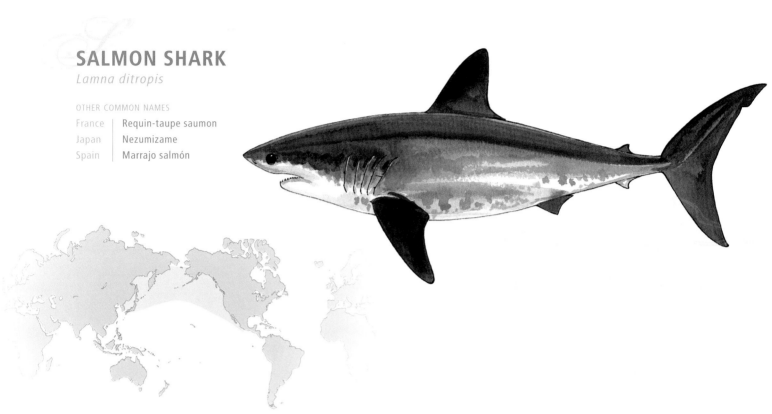

The salmon shark is an important apex predator of northern Atlantic subarctic waters. Its success in this niche is due at least in part to its unusual internal physiology which, like other lamnid (mackerel) sharks, enables it to maintain a core body temperature considerably above that of the cold water in which it swims.

The closest relative of the salmon shark is the porbeagle shark, *Lamna nasus*. These two 'sister' species share many anatomical features, but because their geographic ranges are completely separate, the issue of telling them apart rarely arises. For the record, the salmon shark has a more rounded snout than the porbeagle. On the other hand, two other lamnids – white sharks and makos – overlap with the salmon shark, so some care needs to be taken in separating these species. The best feature to note in the salmon shark is a secondary small lateral keel below the main keel on the caudal peduncle (tail wrist). The white shark and mako lack this smaller keel. In color, the salmon shark is dark gray to black on the dorsal surface, while the white shark is a lighter gray and the mako is blue.

The salmon shark occurs only in the northern Pacific, extending right across between north America and Japan, where it tends to remain for most of the time north of latitude 30°. Its range extends into boreal regions in waters as cool as 2°C – the coldest habitat of any of the lamnid sharks.

The salmon shark has been the subject of intensive research in recent years. Popup satellite tags have been attached to many sharks, mainly off Alaska, and results have shown some remarkable behavior. Female sharks tend to undertake an annual winter trek as far south as the waters off southern California, even as far as Baja Mexico and to within 500 miles of the Hawaiian archipelago. Ambient water temperatures during such migrations range from 2°C to 24°C, demonstrating a very broad range of tolerance for a shark species. This is not the case for all sharks tagged, however, as some stay in Alaskan waters year-round.

The most recent study of the growth rate of salmon sharks was undertaken in the eastern Pacific. This suggests that, at least in this area, salmon sharks have considerably faster growth rates than the same species in the western Pacific, and much faster growth rates than their sister species, the porbeagle shark.

Litter size is two to five pups and breeding may occur only every other year, although females of the closest relative of the salmon shark, the porbeagle, almost certainly produce annual litters. In the eastern Pacific, it is estimated that salmon sharks mature at six to nine years old, while males mature at the young age of three to five years.

A question that emerged while studying the physiology of salmon sharks was: how does their heart muscle cope with the very cold temperatures in which they often remain year-round? As noted, like the other lamnid sharks, the salmon shark is to some extent, warm-blooded. In fact, it has evolved this thermoregulatory ability to a particularly high degree, able to maintain its core temperature at up to 14°C above that of surrounding water. This is achieved by a countercurrent heat exchange circulatory system, which keeps muscles, eyes and brain warm, thereby bestowing an advantage of the salmon shark over its more sluggish cold-blooded prey species. However, the heart is not warmed in this way since it constantly has to pump cold blood coming from the gills. The secret was found to lie in the 'over expression' of two proteins in heart muscle of salmon sharks which enhance transport of calcium within the muscle fibres, in turn, increasing the efficiency of contraction and relaxation of the heart.

In the 1990s, a fishery developed on salmon sharks in Prince William Sound, Alaska. Concentrations of sharks were so large that it was even profitable to purse seine them. Fortunately for the sharks, the Alaskan Government banned commercial fishing for sharks in 1997 and introduced strong regulations on the sport fishery.

THRESHER SHARKS
Alopias vulpinus, A. pelagicus,
A. superciliosus

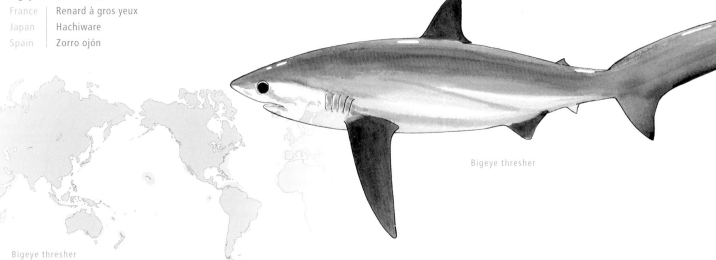

Bigeye thresher

Bigeye thresher

Pelagic thresher

Common thresher

Common thresher

There are three species of thresher shark, the common thresher, *Alopias vulpinus*, the bigeye thresher, *Alopias superciliosus*, and the pelagic thresher, *Alopias pelagicus*. A genetic study in 1995 suggested that there may be a fourth species, very similar in outward appearance to the common thresher, occurring off the Pacific coast of Mexico. However, this finding does not appear to have been supported by any further studies.

Because of their robust body form and apparent highly evolved internal biology, the threshers have often been allied with the lamnid or mackerel sharks – the mako, white and porbeagle sharks. However, threshers lack many lamnid features, such as broad keels at the base of the tail and large gill slits. A relatively recent study of the genetics of these sharks has produced fairly strong evidence that threshers are much more closely related to the sand tiger sharks (family Odontapsidae), and possibly to the large, deep-water megamouth shark.

The upper lobe of the thresher's tail is generally as long as the body of the shark itself – easily the longest tail of any shark species. Within the family, the species can be distinguished as follows. The bigeye thresher has extremely large, oval or inverted pear-shaped eyes, which extend onto the top of the head. It also has a characteristic helmet-shaped bulge of muscle behind the head, demarcated by horizontal grooves on either side above the gills. The head of the common thresher is arched between the eyes, and it is the only species which has grooves at the corner of its mouth (labial furrows). The pelagic thresher also has an arched head between the eyes, but lacks labial furrows. Another distinguishing feature is that the common thresher has a white patch extending from its undersurface forward over the pectoral fin, whereas the pelagic thresher has the dark, dorsal coloration in this area.

The three species of thresher shark are all very widespread. The common and bigeye threshers occur throughout the three major oceans of the world, with the bigeye being more confined to the tropics than the common thresher. The latter extends well to the south of Africa, Australia and South America, and in the north, nearly to Alaska in the Pacific, and as far as northern England in the Atlantic. In contrast, the pelagic thresher is confined to the Indo-Pacific. Only the bigeye and common threshers have been recorded from New Zealand, while all three are found in the Hawaiian Isles and off at least southern California and Baja. For all three species, though, their overall distributions are surmised from rather patchy location information.

Very few thresher sharks have been tagged in cooperative tagging programs. On the NOAA shark tagging program operated from Rhode Island, the most recently available figures show that 329 bigeye threshers and 48 common threshers had been tagged. Seven of the bigeyes had been recaptured, with the longest distance moved being 1494 nautical miles into the Atlantic. The two common threshers were recaptured within 90 nautical miles of their release points, one after eight years at liberty.

An aging study of the bigeye thresher off Taiwan estimated that this species lives up to 20 years old. Ages at maturity were estimated at 12 to 13 years for females (3.3 to 3.4 metres overall length) and nine to ten years for males (2.7 to 2.87 metres). Pelagic thresher sharks have also been aged by Taiwanese scientists, and found to be even longer lived – 28 years for females and 17 years for males. Estimated ages at maturity for this species were a little less than bigeye threshers – eight to nine years for females and seven to eight years for males. Maturing at such relatively old ages as these two species indicates very slow generation turnover times and is a 'red light' indicator that threshers are likely to be prone to overfishing.

Maximum sizes of threshers are not well recorded. The maximum lengths given in the scientific literature, bearing in mind that this includes the tail, are as follows: common thresher 7.6 metres, bigeye thresher 4.88 metres and pelagic thresher 3.83 metres; however, the oft-quoted figure of 7.6 metres for the common thresher seems unlikely. This was derived from an old Canadian publication and converts to a neat 25 feet, which is likely to have been an anecdote or a guess rather than an actual measure.

As with records of hammerhead sharks, IGFA records for thresher sharks are lumped under the one general category. The all-tackle record is a 363.8 kg (802 lb) fish caught off New Zealand in 1981, verified from photographs as a bigeye thresher while preparing this book. The next three largest specimens – 348 kg (767 lb), 335 kg (739 lb) and 330.7 kg (729 lb) – were also caught in New Zealand waters and are also likely to have been bigeyes.

Thresher sharks are ovoviviparous, giving birth to usually only two, but occasionally up to four, quite large pups. The development of the young inside the mother is of great interest. Threshers are intrauterine cannibals, meaning that the developing young eat fertilized eggs (that is, their potential brothers and sisters) inside the oviducts of their mother. The eggs are continually produced during gestation, and the voracious young have enormously distended bellies, filled with yolk, prior to their birth. The size of pups at birth ranges from about 95 cm for pelagic threshers to 150 cm for the common thresher. Gestation periods are unknown.

The mouth and teeth of threshers are quite small, indicating a diet of relatively small items. Threshers eat a wide variety of fishes, including lancetfish, herrings, mackerels and even small billfishes. Squid are also a feature of the diet, as are some bottom-dwelling fishes, indicating that threshers cover a broad depth range in feeding.

And now we come to the function of the incredible tail of the thresher. There is no doubt that the tail is used to herd baitfish, and probably squid, not only underwater but also by violently whacking the sea surface. Small groups of threshers have even been observed to cooperate in such activities. It is highly likely that the tail is also used to actually stun or kill prey. This has been observed in the wild, but better proof comes from the fact that thresher sharks are often hooked on long-lines through their tails, presumably as they strike at the bait. Foul-hooking thresher sharks by the tail is also commonly reported by sportfish anglers off southern California.

Historically, there are persistent stories of thresher sharks attacking whales, and while that would seem highly unlikely, the following eyewitness account by Frank T Bullen, written in 1904, makes fascinating reading:

> I have seen the thresher shark attacking a whale at close quarters ... The shark appears to balance himself upon his head in the water, with the whole of his enormous flail-like fluke in the air at the moment of striking; then, when the blow has been delivered there is a quick descent and return, like the lashing of a gigantic whip, while the blows are audible for two miles on a calm day. So heavy are they that strips of blubber are cut by them from the back of the hapless whale four to six inches wide, and two to five feet in length.

Another explanation could be that the thresher shark might have been attempting to stun fish seeking shelter under the whale, accidentally striking the whale in the process.

On the other hand, whales may prey on thresher sharks. Killer whales in New Zealand have been observed catching and feeding on thresher sharks on at least three separate occasions. The sharks were estimated to be quite small, at 1.5 metres, 2 metres and 3 metres total length. The whales were observed to chase the threshers, catch them, and in one instance, were seen to share parts of the carcass with other adults and calves.

In the 1980s, pioneering work conducted by Frank Carey of Woods Hole Oceanographic Institute revealed that muscle temperatures of bigeye thresher sharks were up to 4.3°C higher than the ambient water temperature. More recently, researchers in southern California have found that the common thresher is much more advanced in controlling its internal temperature compared with the

bigeye and pelagic threshers. The common thresher achieves this by means of an intricate countercurrent blood flow heat exchange system which is extraordinarily similar to equivalent systems in tunas and lamnid sharks (especially porbeagles). Intriguingly, all three groups of fish have evolved this type of heat exchange system entirely independently.

It has also been found that the bigeye thresher has a brain and eye heating organ not unlike that of the broadbill swordfish, but again evolved independently. Such adaptations very likely give these predators a decided advantage when hunting cold-blooded prey in deep water.

Acoustic and popup satellite tagging of common thresher sharks has shown that, while they are generally creatures of the surface at night, they regularly dive down below the thermocline during the day into quite cold water. Dietary studies have confirmed that favorite food items includes hake, which normally occurs at deeper, colder depths. Japanese scientists have also tracked bigeye thresher sharks in the eastern Pacific. The sharks stayed

at depths of between 200 and 500 metres during the day, ascending to within 130 to 80 metres of the surface at night. The deepest dive observed was 723 metres.

Thresher sharks are targeted in some areas, especially in the eastern Atlantic by Spanish longline fleets. Off the Pacific coast of the United States, threshers were historically caught in large numbers by drift net, increasing to about 1000 tonnes per year by 1981. It was not long before signs of overfishing were noticed, including a decrease in average size caught and a marked drop in the catch rate. Subsequent concerns about the sustainability of the thresher population, as well as for turtles and marine mammals, led in 1990 to exclusion of drift nets out to 75 nautical miles of the coast. Fortunately, the thresher population appears to have made a slow recovery, and in doing so, has presented us with some important lessons. Slow-growing, late-maturing sharks do not respond well to heavy fishing pressure. Releasing pregnant female sharks is now being recommended throughout the sport fishery, which makes good sense all round.

A common thresher shark showing its extraordinary tail, as long as its body. All three thresher species use their tails to stun prey.
marinethemes.com/ Kelvin Aitken

HAMMERHEAD SHARKS

Sphyrna mokarran,
S. lewini, S. zygaena

Great hammerhead *S. mokarran*

| Japan | Hira-shumokuzame |
| Spain | Tiburón martillo |

Scalloped hammerhead *S. lewini*

| Japan | Aka shumokuzame |
| Spain | Cornuda común |

Smooth hammerhead *S. zygaena*

| Japan | Shiro-shumokuzame |
| Spain | Tollo cruz |

The hammerhead sharks are among the most distinctive fishes of the world. The sides of the flattened head are extended laterally with the eyes positioned right at the ends. Otherwise, their bodies are typically shark-like, with relatively small mouths, and large dorsal fins originating ahead of the origin of the pectoral fin.

The question most often asked is why do hammerheads have such a bizarre modification to their heads? As usual, scientific opinion leans towards a combination of likely reasons. Firstly, the flattened shape probably acts like a hydrofoil, increasing lift and maneuverability (bearing in mind that sharks are negatively buoyant and need to generate lift as they swim). In fact, the head of hammerheads is sometimes termed a 'cephalofoil'. Secondly, the broad head separates the eyes, perhaps making binocular vision more efficient (although hammerhead sharks often seem to have great difficulty in finding baits). Thirdly, the many sensory capabilities of the shark, such as smell (the nostrils are wide apart) and chemical, electrical and pressure senses, are increased by spreading receptors along the leading edge. This seems a highly likely function of the broad head since hammerheads move their heads rapidly in sweeping arcs when hunting – not unlike biological metal detectors. A fourth possible function is to act as a pinning tool. Hammerheads are highly electro-receptive, being able to detect the presence of fish buried in sand by the tiny electrical fields generated by their muscles. The great hammerhead is particularly fond of stingrays, and has often been observed swimming close to sandy bottoms obviously hunting for this prey. When a buried ray is detected via its weak electric field, the hammerhead flushes it out, then pins it to the substrate with its broad head, while twisting to bite off chunks from the ray's flaps. In fact, it is not unusual to find multiple stingray barbs in the mouths of great hammerheads. In a paper published in 1947, EW Gudger recorded his observations on a 12 foot 6 inch (3.8 metre) hammerhead shark, harpooned off Florida, as follows:

When dissected I found in the stomach an almost perfect skeleton of a stingray with many like fragments of other skeletons, and I got from its throat, mouth and jaws 54 stings, varying from perfect spines to broken-off tips – souvenirs of at least that many stingrays caught and probably eaten. But for all these accumulated stings, this shark was a living dynamo of energy when harpooned.

Gudger called this shark *Sphyrna zygaena* (the smooth hammerhead), but a photo of the shark in his paper clearly shows it to be a great hammerhead (*Sphyrna mokarran*).

There are nine species in the hammerhead family, Sphyrnidae, but in the context of open-ocean fishes, we will concentrate here on the three largest species of hammerhead which also happen to have the broadest geographical ranges. These are the smooth, scalloped and great hammerheads, all of which are relatively easy to distinguish. The smooth hammerhead has a clean, curved line to the margin of its head, with no central indentation. The scalloped has a marked indentation at the midpoint of its head margin, while the great hammerhead has an almost straight anterior edge to its head margin. The great hammerhead also has an extremely high, pointed dorsal fin.

The scalloped, smooth and great hammerheads all occur on both sides of the three major world oceans, in the Mediterranean and around major mid-ocean island groups. The great and scalloped hammerheads have essentially similar tropical to subtropical distributions. The smooth hammerhead has a somewhat more temperate distribution, occurring around the southern parts of Africa, South America and Australia, and is the only hammerhead found in New Zealand. In the northern hemisphere, it occurs in the Mediterranean, and as far north as southern England, Nova Scotia, northern California and northern Japan.

Little work has been carried out on movements of hammerhead sharks. Until 2008, just under 5000 hammerhead

Great hammerhead

Scalloped hammerhead

Smooth hammerhead

sharks (unknown species composition) had been tagged under the Australian Gamefish Tagging Program, but only 54 recaptures had been recorded. All recorded movements had been coastal, with the maximum distance moved being about 300 nautical miles (550 km).

At certain times of the year, the scalloped hammerhead has been observed to form large aggregations of up to 500 individuals either off coastlines (probably to give birth) or above offshore seamounts, one well-known example being in the Sea of Cortez. These aggregations occur annually, in a predictable fashion, and it has been suggested that the sharks navigate to these regions by following magnetic lines of force on the ocean floor. This fascinating theory is yet to be fully proven, however.

It is generally believed that the great hammerhead is the largest of the hammerheads, credited with a maximum length of 5.5 to 6 metres, possibly more. Maximum lengths quoted for other species are: scalloped hammerhead, 4.2 metres, and smooth hammerhead, 4 metres.

The all-tackle world record hammerhead weighed 580.59 kg (1280 lb) and was caught off Florida in 2006. This surpassed the previous record of 449.5 kg (989 lb) – also caught off Florida, in 1982 – by a substantial margin. The next largest was a 358 kg (788 lb) specimen caught off New South Wales, Australia, in 1999. The largest smooth hammerhead recorded by IGFA weighed 164.6 kg (363 lb), caught off the Azores, and the largest scalloped hammerhead, 160.1 kg (353 lb), caught off Florida.

Of the three species considered here, the growth rate of only one has been examined in any detail – the scalloped hammerhead. Several studies, conducted in the Gulf of Mexico/northwest Atlantic, Taiwan and the Pacific Mexican coasts, have resulted in growth estimates

for this species. The Atlantic study estimated growth rate to be very slow, with fish only reaching about 120 cm total length by five years of age, and by age 15, still not attaining 200 cm. This was in contrast to studies in the Pacific which suggested a much faster growth rate. The main reason for the discrepancy was the assignment of two vertebral growth bands per year in some studies, but only one in others. This issue has still not been satisfactorily resolved.

Female scalloped hammerheads attain maturity at a much larger size than males (about 230 cm and 175 cm respectively). As with all the other pelagic sharks, hammerheads give birth to live young. Hammerheads are ovoviviparous, with the young developing inside the mother, but not connected to her by means of a placenta. The number of young is variable within and between species. Great hammerheads produce between 20 and 42 pups per litter. The gestation period is estimated at about seven months, with each pup measuring between about 50 and 70 cm at full term. The scalloped hammerhead gives

birth to between 15 and 30 pups, at a size of 40 to 50 cm, while smooth hammerheads produce 30 to 37 pups, which measure 50 to 61 cm at full term. The scalloped hammerhead moves into shallow water to give birth. In Hawaii, this occurs in shallow lagoons, and off the southeastern Australian coast, it occurs just outside the ocean surf zone, and possibly inside the mouths of deeper estuaries.

Hammerhead sharks are active predators but because their mouths and teeth are relatively small, they are not generally credited with the ability to take large prey animals such as marine mammals. Food items found in hammerhead stomachs include pilchards, eels, halfbeaks (garfish), mullet, barracuda, trevallies (jacks), wrasses and other reef fish, bluefish (tailor), flounder, squid, octopus, lobsters and pieces of stingray. Indicating their fondness for stingrays, great hammerheads are often found with multiple stingray spines embedded in their mouths and heads, which always makes human observers wince, but apparently, and surprisingly, not the sharks themselves.

Even though their diets are almost exclusively small items, there are some reports of attacks on humans by very large (presumably great) hammerheads. These derive from Second World War accounts of attacks on survivors of torpedoed ships, and while they may be true, they do not appear to have been completely verified. Divers have occasionally reported aggressive encounters with hammerheads, but more often than not, these generally shy sharks are frightened away by scuba bubbles, rather than attracted to swimming humans.

Aggregations of scalloped hammerheads over seamounts (see above) consist mainly of subadult females which constantly interact with each other in trying to take up central positions in the group.

The reason for such behavior is not clear, although it has also been observed that fully adult females also attend the seamounts where they mate with males. Interestingly, the sharks only aggregate near the seamounts during the day, but disperse at night, presumably to forage.

Hammerhead sharks are not specifically targeted by commercial fisheries in many parts of the world, although in Southeast Asia, they form a significant component of the inshore gillnet fishery. They are also a consistent component of the bycatch of longline fisheries, especially where these occur on or near continental shelves. Recently, the scalloped and great hammerheads were listed as 'vulnerable' by the International Union for the Conservation of Nature.

WHALER SHARKS
Carcharhinus spp.

OTHER COMMON NAMES

Requiem sharks

Carcharhinids

The term 'whaler' shark is used here to include only members of the genus *Carcharhinus*. The term 'requiem' shark is also sometimes used in the same way, but is more often applied to all 52 species within the family Carcharhinidae. This family not only includes the whalers, as here defined, but also the tiger shark, blue shark (both dealt with elsewhere in this book) and several other genera of ground sharks including the sharpnose, sliteye and speartooth sharks.

The name 'requiem' shark is almost certainly derived from the generic French word for shark – 'requin', but this in turn is likely to be derived from the original Latin 'requiem', which means 'to rest', or in its religious context, 'requiem in pace' or 'rest in peace' – a reference no doubt to the reputation of sharks as killers. The origin of the name 'whaler' shark is also somewhat obscure. Some suggest that it was derived from the habit of these sharks in following whaling vessels and attacking whale carcasses after they were killed. This is hard to verify, but may well be correct.

The whaler sharks *look* like sharks. All of the many closely related species of the genus *Carcharhinus*, ranging from the small tropical reef whalers to the pelagic oceanic whitetip, have the classic, almost generic body form of the sleek, fast-swimming predator we all associate with the word 'shark'.

Worldwide, there are at least 29 species of whaler sharks. Although they are all closely related, the genus is quite diverse, having successfully radiated into a wide variety of habitats including the open ocean, continental shelves, coastal environments, estuaries and even fresh water. Some species grow no larger than 1 metre, while others may approach 4 metres in total length.

Most of the whaler sharks have a highly evolved form of reproduction, termed 'viviparity'. This means that the embryos develop inside the mother, and are connected to her by a functional placenta, very much like mammalian fetuses. The shark placenta is actually a highly modified yolk sac which is well supplied with blood vessels and acts functionally like a true mammalian placenta. The young are born as well-developed miniatures of the adults, often quite large in relation to the size of the mother, compared with many other shark species.

As a generalization, whaler sharks are very slow-growing. We know this from tagging programs in which some species of whaler shark have been recaptured after many years (see the sandbar shark below) and accurate growth estimates can be made. And while slow growth rates of whaler sharks appear to be the norm, some species grow more rapidly. For example, the Australian blacktip shark (*Carcharhinus tilstoni*) grows from a size of 60 cm at birth to about 115 cm by an age of three to four years. These small whaler sharks only grow to a maximum size of about 50 kg (110 lb), at which size they are estimated to be about 12 years old.

The five species of whaler shark selected below for fuller treatment are either oceanic or are at least very widespread, occurring in all three major oceans. All five grow to a large size and are important apex predators in their respective habitats.

Oceanic whitetip shark
Carcharhinus longimanus

This impressive and most distinctive member of the whaler family is also the most adapted to oceanic waters. The oceanic whitetip is an active predator of the surface layer of tropical and subtropical oceans. It has an extremely broad distribution throughout the Pacific, Indian and Atlantic oceans. In the Indo-Pacific, it is found between the latitudes of about 30°N and 30°S (including both eastern and western Australia), whereas in the Atlantic its range extends beyond 40° in both hemispheres.

The oceanic whitetip is one whaler shark that is easy to identify. It has a very rounded, high first dorsal fin, marked with a white tip which mottles into the gray of the body about a third of the way down. Its long pectoral fins also have markedly rounded tips, with similar terminal white markings. The purpose of this distinctive color pattern is not clear, but it has been speculated that the three diffuse white patches resemble small schools of baitfish. These may attract other predatory fish such as tuna and marlin which are pounced on by the shark. This idea might seem somewhat fanciful, but it is known that larger fish, including marlin, do form part of the diet of whitetips, so there may be some basis to the theory.

The oceanic whitetip is not generally thought of as a large shark, and while individuals over 350 kg (770 lb) have been reported (and I have seen one at least that large), the IGFA all-tackle record stands at only 167 kg (369 lb). The number of pups in a litter ranges from one to 15, measuring about 60 cm at birth. Some limited tagging of oceanic whitetips off the eastern United States has resulted in some information on movements. Only 540 taggings have been recorded, and of only six recaptures, the furthest distance moved was an impressive 1220 nautical miles (2260 km) by a whitetip tagged in the mid-Atlantic.

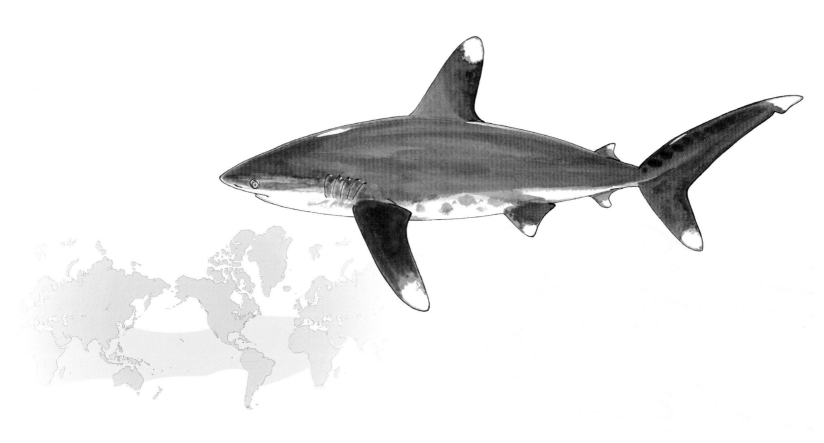

The most oceanic of the whaler sharks, the oceanic whitetip is an efficient apex predator of the surface. Pilotfish are nearly always seen in close attendance to oceanic whitetip sharks.
James Watt/ imagequestmarine.com

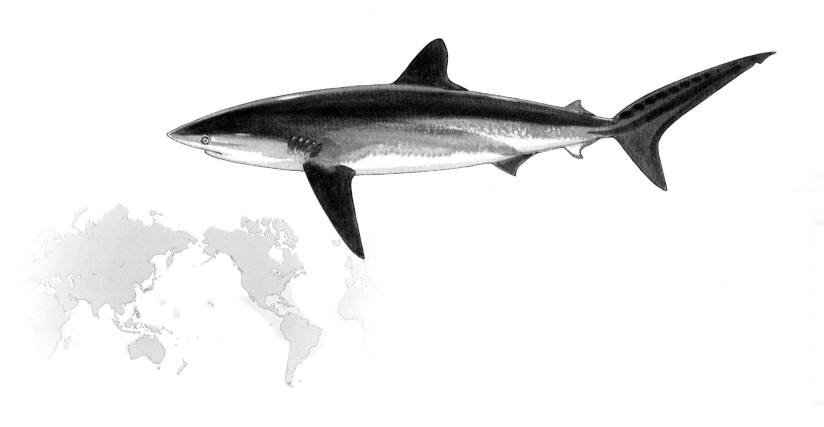

Silky shark
Carcharhinus falciformis

The silky shark is regarded as a true open-ocean species, but is encountered close to shore as well. Like the similar dusky shark, it is relatively plain colored (brown to gray) and has a ridge between the dorsal fins, but may be identified by its second dorsal and anal fins. The loose, trailing edge of these fins is at least twice as long as the height of the fins, a good, reliable character to look for. As the name suggests, the skin of the silky feels very smooth when stroked towards the tail, but this is also true for some other whalers.

The silky shark has a broader range than the dusky, extending to island groups in all oceans, including, in the Pacific, New Caledonia, Kiribati, New Zealand and Hawaii. This species may grow to a large size, the IGFA record being 346 kg (763 lb), caught off Sydney, Australia in 1994. Silky pups usually number between two and 14 per litter and are born smaller than duskies – about 70 to 85 cm. While silky sharks have no doubt been tagged on the Australian Gamefish Tagging Program, because whalers are not identified to species level, we need to look at the US program for any information on movements. On that program, over 800 have been tagged and 54 recaptured. The longest distance moved by a tagged silky shark is 720 nautical miles (1300 km) and longest time at liberty, 7.1 years. Most movements of tagged silky sharks have been coastal, although there have been several offshore movements as well.

Dusky shark
Carcharhinus obscurus

One of the most common of the whalers is the dusky shark, also, not surprisingly, known as the common whaler. In some parts of its range, the dusky is probably quite often misidentified as the 'bronze' whaler. The name 'bronze whaler' is commonly used, and while there is indeed a true bronze whaler species (*Carcharhinus brachyurus*), also called the copper shark, several other whaler sharks may also exhibit a coppery sheen, with the result that a number of whaler species are wrongly labelled with this term. The copper and dusky sharks may be distinguished by a marked ridge which runs along the top of the back between the two dorsal fins in the dusky, but is nearly always absent in the bronze or copper shark. However, for a definitive identification of these, and most other whaler sharks for that matter, the teeth in the upper jaw need to be inspected. In the case of the copper shark, the teeth are relatively narrow, with a marked notch on the outside margin, while those of the dusky shark are broad, and almost triangular.

The dusky shark has a widespread, but patchy distribution. It occurs in all three major oceans, mostly inside the continental shelves of the major land masses and some large islands. Litter size in dusky sharks is between three and 14, and pups are quite large when born – up to 100 cm long. This is a slow-growing shark, taking up to 18 years to reach a length of about 280 cm. The largest dusky whaler positively identified as such weighed 323.5 kg (712 lb) and measured 345 cm, so it is likely that a shark of that size would be at least 30 years old.

As noted above, in Australia, whaler sharks have been tagged by gamefish anglers in quite small numbers, but they are rarely, if ever, identified to species level when released or recaptured. This means that any movements of recaptured whaler sharks are simply designated 'whaler', with little meaningful biological information on each species accruing as a result. However, off the northeastern United States there are fewer confusing species than in Australian waters, resulting in much better identification by anglers. This has meant that several species of whaler shark have been tagged in quite large numbers by recreational anglers in that part of the world. The dusky shark is one of these, with nearly 6000 tagged and more than 120 recaptured. The results have shown almost exclusively coastal movements of this species, along the entire eastern seaboard of the United States. Quite a few of these movements have exceeded 1000 nautical miles, with the record being just over 2000 nautical miles. Remarkably, the longest time at liberty for a tagged dusky shark has been 15.8 years, second only to the sandbar shark.

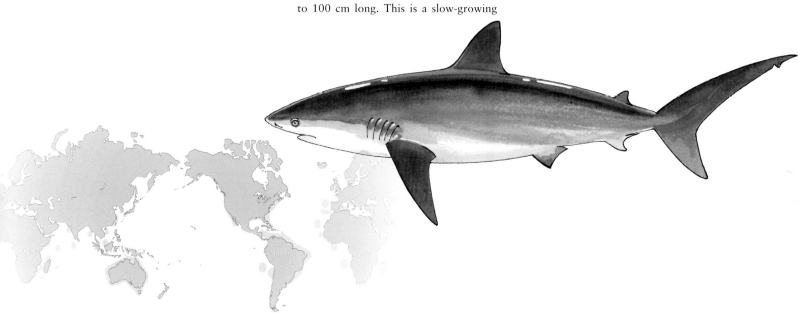

Sandbar shark
Carcharhinus plumbeus

The sandbar is a common species of whaler with a worldwide distribution. It is a distinct-looking shark which is relatively easy to identify. It has triangular teeth, a ridge between the two dorsal fins and a rounded snout, but its chief characteristic feature by far is its very high dorsal fin which is positioned well forward on its body. Technically, the height of the dorsal fin measures up to 16 per cent of the total length of the body, and its anterior origin is at least over, or even in front of the rear insertion of the pectoral fin. This is much further forward than any of the other whalers and gives the sandbar the impression of being 'front heavy'. Sandbar sharks are often a coppery or bronze color, but are readily distinguished from the true bronze whaler or copper shark (*Carcharhinus brachyurus*) by the features of the dorsal fin described above.

Like the other whaler sharks outlined here, the sandbar shark is also an inhabitant of all three major oceans, with a similar distribution to the silky shark. It occurs through tropical Australia, around New Caledonia and Hawaii, but eastern Pacific records off Mexico and the Galapagos Islands are suspect. It is common along the US eastern seaboard, Mexico and Peru and also occurs throughout the Mediterranean.

The sandbar shark grows to a maximum size of about 3 metres and perhaps 250 kg (550 lb), with the IGFA record being 240 kg (529 lb). As for all the whalers mentioned here, sandbars have a yolksac placenta, and give birth to large live young numbering between one and 14 and measuring 42 to 75 cm. Sandbar sharks have been tagged in large numbers off the northeastern United States, second only to blue sharks. Over 15 000 releases have resulted in 720 recaptures. These indicate quite extensive seasonal movements, from north to south of over 2000 nautical miles (3700 km). Interestingly, only a few of these recaptures have been recorded away from the coast.

The longest time at liberty for a tagged sandbar shark is an extraordinary 27.8 years. Growth studies have confirmed that this is a slow-growing species, with females maturing at an average age of 16 years and males at 14.

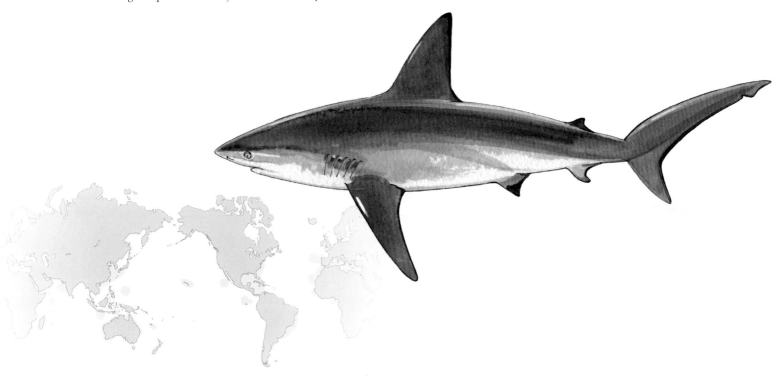

The sandbar shark is one of the most distinctive of the whaler group, readily identifiable by its high dorsal fin, located well forward on the body.
James Watt/
imagequestmarine.com

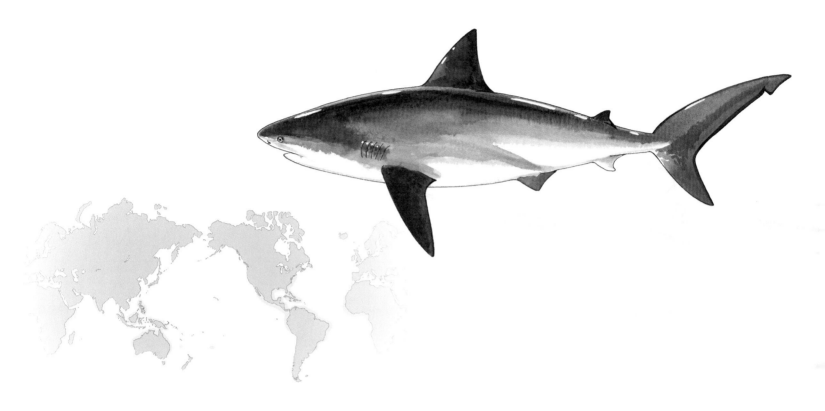

Bull shark *Carcharhinus leucas*

The bull shark is a particularly heavy-bodied shark with a short, relatively blunt snout. It is quite plain colored, although juveniles may have dusky fin-tips. It lacks a ridge between the two dorsal fins, which immediately separates it from the sandbar, dusky and silky sharks. The teeth in the upper jaw of the bull shark are large, serrated and nearly triangular in shape.

The primary habitat of the bull shark is along coastal regions, especially near the mouths of large rivers. It is also perhaps the largest shark which spends significant time inside estuaries, often penetrating into fresh water, especially when juvenile. This species has been implicated in many attacks on humans and is regarded as one of the three most dangerous shark species, along with the white shark and the tiger shark.

The bull shark has a global distribution in tropical and temperate waters. As noted, it is primarily a coastal species, but also occurs around oceanic islands including, in the Pacific, New Caledo-nia, Fiji and the Cook Islands, and in the Atlantic, through the Caribbean and the Bahamas.

The bull shark grows to at least 3.4 metres in length, possibly up to 4 metres. The IGFA all-tackle record for the bull shark is a 316.5 kg (698 lb) specimen caught at Malindi, Kenya, in 2001.

Litter sizes range up to about 13. The gestation period is ten to 11 months and the young are born at 55 to 80 cm long.

Studies in South Africa strongly suggest that bull sharks are very slow-growing and late-maturing. The researchers examined growth rings on vertebrae from wild sharks, and considered growth rates of captive specimens which had been in captivity for up to 20 years. They concluded that the smallest mature female shark sampled was 14 years old and the smallest mature male, 25 years old. The oldest sharks aged were estimated at 32 years for a female shark weighing 238 kg (523 lb) and 29 years for a male shark weighing 180 kg (396 lb).

Whaler sharks tend to be particu-larly good eating, with white, boneless flesh and little of the ammonia taint which other sharks often acquire. This attribute has meant that whaler sharks of many species are the targets of commer-cial and artisanal fisheries in many parts of the world, and because they are usu-ally such slow-growing species, concerns are rightly held regarding their ability to sustain such increasing fishing pressure.

Whaler sharks are often targeted, but are also caught quite commonly as bycatch in various large-scale commer-cial fisheries. Many thousands of silky, dusky and oceanic whitetip sharks are caught and discarded each year by the open-ocean longline and purse seine fleets of the world. As our knowledge of these important sharks, and their roles in marine ecosystems, slowly expands, such collateral mortality is coming increasingly under scrutiny. It is to be hoped that measures taken to mini-mize such wastage are forthcoming, are implemented, and most importantly, are successful.

The bull shark is
mainly an inshore
species that hunts
fish. Its attacks
on humans are
thought to be
random and
accidental.
*marinethemes.com/
Mark Conlin*

TIGER SHARK
Galeocerdo cuvier

OTHER COMMON NAMES

France	Requin tigre commun
Hawaii	Mano pa'ele
Japan	Itachizame
Spain	Tintotera tigre

With its tendency to prowl coastal waters, enter estuaries and follow sailing ships, it is highly likely that at least some of the first known shark attacks were perpetrated by tiger sharks. The frightening habits of (presumably) tiger sharks are well recounted in the following excerpt from the writings of British zoologist Thomas Pennant, who wrote in 1768:

> This [shark] grows to a very great bulk ... to a weight of 4000 lbs and in the belly of one was found a human corps entire, which is far from incredible, considering their vast greediness after human flesh. They are the dread of sailors in all hot climates, where they constantly attend the ships in expectation of what may drop overboard; a man that has that misfortune perishes without redemption.

This passage is often thought to refer to the great white shark, but the scavenging nature of the shark in question, as well as the reference to 'hot climates', point to the culprit being the ubiquitous tiger shark.

The tiger shark has a markedly broad, blunt snout, long furrows leading from the corner of its mouth, a low lateral keel on the tail wrist, large head and shoulders, but slender rear body. As its name suggests, the tiger shark is usually marked with broken vertical stripes which are very prominent when young, but fade with increasing size. Sometimes, these markings are not obvious, occasionally leading to some confusion in identification with the broad-snouted bull shark, *Carcharinus leucas*. The teeth of the tiger are highly characteristic, however, being curved and serrated on the inner side and hooked with larger serrations on the outer side. Bull shark teeth, on the other hand, are almost triangular in shape and lack obvious serrations. Occasionally, small whale sharks (*Rhincodon typus*) may be confused with tiger sharks, at least momentarily. The misidentification is normally quickly realized since whale sharks have several raised ridges running the length of the body plus a pattern of light-colored spots rather than dark, broken stripes. Nevertheless, some of the very large sizes attributed to tiger sharks may well be due to misidentification of whale sharks.

This is a shark with worldwide distribution, tending to be concentrated in the tropics, but extending well into temperate waters. It is largely confined to continental shelves of all continents,

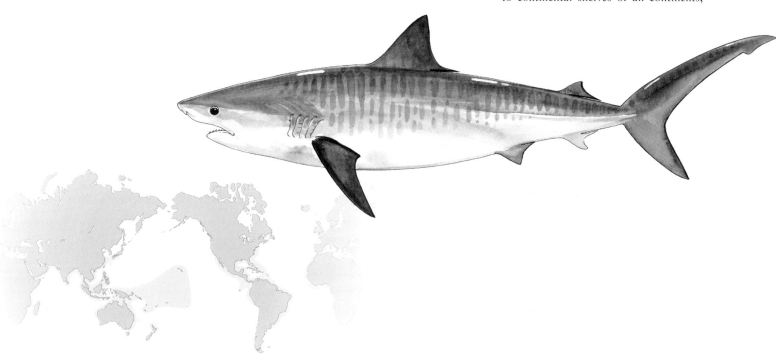

but it is also found around major islands such as Madagascar, Papua New Guinea and New Zealand, and larger archipelagos. It is not regarded as a truly oceanic shark, but is occasionally caught on the high seas and its occurrence near isolated islands suggests that it does at least occasionally travel across open seas. This is also supported by genetic studies which so far have not revealed any significant differences among widely separated populations of tiger sharks.

The only area in which tiger sharks have been tagged in large numbers is off the eastern United States, under the National Marine Fisheries Service (NMFS) cooperative tagging program operated from the Rhode Island laboratory. Over many years, 4850 tiger sharks have been tagged on this program, and 446 recaptures reported. Most tiger sharks tagged have been released from the east coast of Florida. Lengthy coastal movements from there, as far north as the Canadian border, have been recorded, while offshore movements to Caribbean islands and the coast of Venezuela on the South American continent have demonstrated occasional large-scale displacements of tagged sharks. However, movements greater than 1000 nautical miles are the exception to the rule, and to

date, no transatlantic crossings have been recorded. A recent study in Australia, using popup satellite tags recorded an oceanic movement of a 300 kg (660 lb) female tiger shark from the east coast to New Caledonia, a distance of 1800 km (970 nautical miles).

Growth rates of tiger sharks have been studied in a number of different locales, and these have shown some consistency in their results. Such studies rely on counting growth rings on vertebrae, and making the assumption that a new ring is laid down once a year. Using this method, it has been estimated that tiger sharks reach about 2 metres or 70 kg (150 lb) after three years, about 2.5 metres or 160 kg (350 lb) after five years, and 3.5 metres or 400 plus kg (880 lb) by 13 years or so. The maximum life span of tiger sharks, given that they attain at least 800 kg (1760 lb) in weight, is thought to be at least 50 years.

The maximum size to which the tiger shark grows is a point of considerable conjecture. A figure of 9.1 metres total length is often cited, but without actual confirmation, this may be rather fanciful. It would seem that the source of this oft-quoted length is most likely a 1937 publication which stated that the tiger shark 'grows to a length of 15–20 feet, occasionally reaching 30 feet'. Thirty feet converts to 9.1 metres, which then sounds like an actual measurement, but in all likelihood, the figure derives from the above statement. There is also one recurring reference to a tiger shark of 7.4 metres, weighing 3110 kg (6840 lb), being caught in southeast China in 1957, and while this is not possible to prove or disprove, this size seems questionable. In fact, the weight of this shark is cited by various authors as both 3110 kilograms and 3110 pounds. The latter converts to 1414 kg, which is probably more realistic. The maximum scientifically verified length for a tiger shark is 5.5 metres (18 feet) for a shark caught in Cuba.

The all-tackle world record for tiger sharks was a huge specimen weighing 810 kg (1782 lb) caught off New South Wales, Australia, in 2004. Note how much smaller this is compared with the reputed specimen weighing over 3000 kg (6600 lb), again pointing to the latter as being very doubtful.

Developing tiger sharks hatch from eggs inside the mother's body, without any form of placental attachment. This mode of reproduction is considered to be relatively primitive, in terms of the evolution of live bearing. The number of tiger shark pups produced per litter varies considerably, with a range of ten to 82 being most commonly cited. The usual number of pups ranges between 30 and 50. Tiger sharks are born at 50 to 75 cm long, which is quite small for live bearing sharks. The gestation period of the tiger shark is estimated at 13 to 16 months. Mating is thought to occur in spring in both northern and southern hemispheres, with young therefore being born the following summer.

Of all sharks, the tiger has justifiably the strongest reputation as a scavenger, eating virtually anything it comes across. The list of dietary items found inside tiger sharks is long, and includes many species of fish, sharks and rays, lobsters, crabs, squid, marine mammals of all kinds and even jellyfish. But even though tiger sharks appear to be virtually omnivorous, the prey which they seem to especially hunt are sea turtles, sea snakes

Full-term baby tiger shark taken from its mother after capture. Note its relatively small size and the highly contrasting body striping which becomes less apparent with age.
Julian Pepperell

and various sea birds, especially shear-waters. Remains of sea turtles are commonly found in tiger shark stomachs, and it is likely that this shark is the only predator capable of biting through the hard shells of large turtles. Marine mammals, especially seals and dolphins, but also dugongs and whale parts, are also consistently found in tiger shark stomachs. This sometimes surprises those who consider the tiger shark to be a sluggish swimmer, but it is capable of rapid bursts of speed, and also may ambush small dolphins and dugongs at night, although this has not been directly observed.

Tiger sharks often congregate near the mouths of rivers and estuaries, especially after floods, consuming many drowned animals including chickens, pigs, dogs, sheep, cattle monkeys, and humans. As well as organic material, tiger sharks are also known to eat an incredible array of inanimate objects, resulting in some humorous descriptions of the contents of tiger shark stomachs such as that of David Stead, an early

Australian ichthyologist, who wrote in 1933: 'A 10-foot specimen ... had in its stomach, a woman's handbag in which was a comb, a pencil, a powder puff, and a wristlet watch (which was still going!)'.

Tiger sharks are regarded as probably the most dangerous shark in the world with respect to attacks on living humans. The other candidates are the great white and bull sharks; however, neither of these occupies as large an area and as wide an array of niches as does the tiger, so in all likelihood, the tiger shark has been responsible for more attacks on humans than any other species.

Being a large, relatively common shark, the tiger is often taken in commercial fisheries for its fins, which bring a high price on Asian markets. The flesh is regarded as inedible in some areas but is used for human consumption in many others. Recreational shark fisheries target tiger sharks in some areas, notably off Florida in the United States and off southeastern Australia.

BLUE SHARK
Prionace glauca

OTHER COMMON NAMES

	Blue whaler
	Blue dog
France	Peau bleue
Japan	Yoshikirizame
Spain	Tiburón azul

The blue shark is quite probably the most ubiquitous shark in the world, certainly in the mixed layer of the open ocean. It is a fecund, relatively fast-growing shark which plays an important role in oceanic ecology. It is caught in large numbers in many oceanic surface fisheries, either intentionally or as bycatch, and is also an important sportfish in colder latitudes.

This distinctive shark is identified by its long, relatively pointed snout, its very long, floppy pectoral fins (as long as the head) and its distinct deep, almost metallic blue coloration on the back and upper flanks, contrasting with the stark white undersurface. The body of the blue shark is noticeably long and slender, lending it a sinuous, eel-like motion when swimming. The upper lobe of its tail is about twice the length of the lower lobe and the caudal peduncle (tail wrist) has no lateral keels. The only other sharks with which the blue shark might possibly be confused are the shortfin and longfin makos. However, the former has a robust build, while both makos have a nearly symmetrical tail and large lateral caudal keels.

The blue shark is a member of the family Carcharhinidae, which also contains the requiem or whaler sharks. However, the blue is somewhat of an outlier within this family. The main feature that distinguishes the blue from the whaler sharks is a technical one – the position of the first (main) dorsal fin. In the blue shark, the midpoint of the dorsal is closer to the paired ventral or pelvic fins than to the origin of the pectoral fins, while in the whalers (all of which belong to the genus *Carcharhinus*) the reverse is the case.

The blue shark is a truly cosmopolitan species and one of the most widespread species of shark in the world. It ranges across all three major oceans, between latitudes as high as 60°N and 55°S. It is generally described as a cool water shark, preferring temperatures between 7°C and 16°C. In the tropics, blue sharks tend to stay below the thermocline, at depths of between 80 and 100 metres, where they are vulnerable to longline gear. As well, they tend to be part of pelagic ecosystems which associate with floating logs, thereby being vulnerable to purse seining.

Of all of the oceanic sharks, the blue has been tagged in by far the largest numbers, and consequently, more is known about its movements than any other shark. In fact, with over 95 000 tagged in the Atlantic, mainly off the eastern United States and more than 5800 recaptures,

more is probably known about movements of the blue shark than any other oceanic fish. The results clearly show that blue sharks are extremely mobile, regularly crossing the Atlantic and making some trans-equatorial movements as well. In fact, a map showing all of the recaptures looks like a random scatter of movements from release areas to points all over the northern Atlantic.

Recaptures of blue sharks tagged off eastern Australia have not shown quite such extensive movements as those tagged off the United States, although quite a few long-distance movements into the southwestern Pacific have been recorded. As well, two blue sharks, one tagged off Victoria, and one tagged off Tasmania, were recaptured off Java in the Indian Ocean, distances of over 4000 km. While no crossings of the Pacific or Indian oceans have been recorded, this may

simply be because the numbers tagged in those oceans compared with the Atlantic are small. One blue shark did come close to crossing the Pacific, however. That fish was tagged off New Zealand and recaptured 1200 km short of the coast of Chile. Other blue sharks tagged off New Zealand have been recaptured through the southwestern Pacific.

As with other shark species, the growth rate of blue sharks has been studied by staining vertebra and counting growth bands, and in some cases, comparing results to length frequency analysis. A number of such studies in different parts of the world have produced varying results, but taking the average growth rates derived from these studies, blue sharks reach about 80 cm total length after one year, 150 cm by three to four years old, 200 cm (about 30 kg – 66 lb) by five years of age and about

Possibly the most widespread shark in the world, the blue shark spreads its long pectoral fins and rides the ocean currents.
marinethemes.com/ Kelvin Aitken

280 to 300 cm (100 to 120 kg – 220 to 265 lb) after ten years. It is estimated that blue sharks live for up to 20 years, with the maximum total length recorded being 3.8 metres. The IGFA world record blue shark weighed 239.5 kg (528 lb) and was caught off Montauk Point, New York, in 2001. Another weighing 205 kg (451 lb) was caught off Massachusetts, while two others which weighed just under 200 kg (440 lb) were landed off southeastern Australia and New Zealand.

As is the case with some other shark species, male blue sharks bite females quite deeply during mating. This invariably leaves characteristic bite marks and scars on females, which possess very thick skin compared with males. It is estimated that males mature by four to five years old (about 200 cm total length) and females by five to six years old (185 to 210 cm long). Female blue sharks almost certainly store sperm, fertilizing their eggs up to a year after mating. Gestation takes nine to 12 months, with the number of young varying considerably. The blue shark is one of a number of shark species which has evolved the most advanced form of live bearing, known as viviparity. This term means that the embryonic sharks developing inside their mother actually attach to her by means of a kind of placenta. This organ is not a true placenta, but is derived from the yolk sac of the embryo. There has been

a wide range reported in the number of pups in litters of blue sharks, each measuring 35 to 50 cm, from one to 135, with an average of about 35. (As with counts of pups from other sharks, it is likely that the lowest counts resulted after abortion of near-term embryos upon capture of the mother.) This relatively high fecundity no doubt contributes to the blue shark being so prolific, and also, fortunately, to its apparent ability to withstand overexploitation.

The blue shark feeds on a broad range of prey items, including many species of bony fish, small sharks, squids, crustaceans, especially pelagic crabs, dead marine mammals and occasionally, sea birds. In some areas, blue sharks follow squid in their daily vertical migration, moving deeper during the day, and towards the surface at night. They commonly associate in numbers around floating logs as well as near fish aggregating devices which are placed by purse seine fleets to aggregate tropical tunas.

In one particularly interesting experiment conducted off Hawaii by the US National Marine Fisheries Service, 14 blue sharks were caught on longlines and tagged with popup satellite tags to log their daily diving behavior and movements. All but one of the tags deployed to the surface and transmitted data after periods of one to 238 days. During the day, the sharks spent much of the time

at depth, often swimming below the thermocline to maximum depths of about 400 metres. At night, their behavior changed, spending nearly all their time within 200 metres of the surface or less. About 75 per cent of the sharks' time was spent in water temperatures ranging from 20°C to 25°C, while 90 per cent of their time was spent at temperatures between 11°C and 25°C. There was also some suggestion that females were more likely than males to swim nearer the surface.

While blue sharks are rarely targeted commercially, they are nevertheless taken in very high numbers as bycatch of longline and purse seine fisheries. Until finning of sharks by US longliners was banned in the north Pacific, tens of thousands of blue sharks were finned there. Of course, other fleets also take blue sharks, and in 1996, it was estimated that about 138 000 tonnes of blue shark were taken by longline fisheries in the entire Pacific. If each shark weighed 50 kg (110 lb), this tonnage converts to 2.7 million individual animals. As well, very large numbers of blue sharks are caught by purse seiners setting on logs or artificial fish aggregating devices. One estimate of the total global mortality of blue sharks put the figure at between 10 million and 20 million animals. The effects of removing such large numbers of a key predator from ocean systems are entirely unknown.

CROCODILE SHARK
Pseudocarcharias kamoharai

OTHER COMMON NAMES

France	Requin crocodile
Japan	Mizuwani
Spain	Tiburón cocodrilo

The crocodile shark is an unusual member of the lamniform sharks (mackerel sharks), a group which also includes the great white, mako and salmon sharks, as well as the threshers and sand tigers. However, it is sufficiently different from all of these to have been placed alone in its own family, Pseudocarchariidae.

The Japanese name for this little shark, 'mizuwani', translates literally as 'water crocodile' – not only because it has sharp teeth but also because it snaps its powerful jaws open and shut when caught, sometimes taking longline fishermen unawares.

The crocodile shark has a very low first dorsal fin, long sharp teeth, large eyes, short pectoral fins and an asymmetrical tail. Another good feature to look for is a whitish patch between the mouth and the gills. The liver of the crocodile shark is very large, comprising about 20 per cent of the total weight of the body. Shark livers are filled with oil, so the large liver no doubt provides considerable buoyancy to this pelagic shark.

This is a small shark, only growing to a maximum length of about 1.1 metres. It is firmly muscled and possesses sharp teeth and protrusible jaws – good indicators of an active predatory lifestyle.

The crocodile shark has a somewhat patchy, global distribution, with continental associations but also concentrations in mid-oceanic regions. Recent catch information has shown it to be particularly abundant off the Western Australian coast. While considered a true pelagic shark, it undertakes daily vertical migration. Remaining as deep as 600 metres during the day, it ascends to the surface at night, as part of the so-called deep scattering layer community which includes vast quantities of zooplankton, squid and deep-water fishes. This habit makes it vulnerable to capture by longline, one of relatively few species of shark which is regularly caught by this method.

As is the case with other members of the lamniform sharks, crocodile shark reproduction involves intrauterine oophagy, meaning that the first embryos to develop continually eat eggs that the mother keeps producing. Usually two young develop in each of the uteri, resulting in four full-term pups.

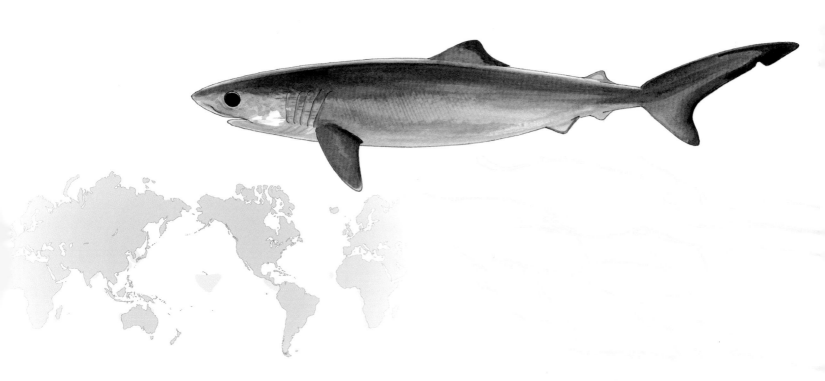

COOKIECUTTER SHARK

Isistius brasiliensis,
I. plutodus

OTHER COMMON NAMES

	Luminous shark
France	Squalelet féroce
Japan	Darumazame
Spain	Tollo cigarro

Most commercial and recreational fishers will never have seen a cookiecutter shark, but many will be well aware of the evidence of their existence – deep, round wounds on the sides of tuna, billfish and other pelagic species. Observations of these wounds, not only on large fish but also on marine mammals, had puzzled scientific minds for many years until it was proven that they were caused by a small shark with unusual teeth – the bottom row are sharp and flat, closely resembling a band saw, while the top teeth are narrow and hook like. Not surprisingly, the shark has been dubbed the cookiecutter.

There are two species of cookiecutter shark, both worldwide – the common cookiecutter, *Isistius brasiliensis*, and the aptly named largetooth cookiecutter, *Isistius plutodus*. Presumably the largetooth has evolved to take advantage of animals with thicker skin, such as larger sharks and whales, although no comparative studies have been undertaken of the diets of the two cookiecutter species.

The geographical range of the cookiecutter sharks has been surmised to some extent by the presence of fresh crater wounds on the flanks of their victims. These show the species to be extensively spread throughout the world's major oceans, albeit patchily.

The question most often asked, of course, is how does the cookiecutter actually manage to take a bite out of much larger, usually predatory, fishes and marine mammals? The answer almost certainly lies in another feature of the shark's biology which gave it its Latin name – *Isistius*, after Isis, the Egyptian goddess of light. In fact, the cookiecutter shark glows in the dark, the eerie light coming from little groups of cells called photophores arranged along the under-surface of the body, with the exception of a dark collar around the neck. Many deep-sea fishes and squids also have glowing undersurfaces and it is likely that the glow pattern of cookiecutters has evolved to conceal them when viewed from below against scattered back lighting. But here is the trick of the cookiecutter. It seems that the dark patch under its head, surrounded by phosphorescence, may act as a lure to upward-looking predators. Then, when they rise to the bait, as it were, the little shark senses them coming, perhaps switches off its lights, and makes its attack. As the startled predator swims off, water pressure is used by the shark to spin and cut a chunk with its lower teeth, completing its parasitic mission in a matter of seconds. Whatever the maneuver, it must be highly effective since the list of species known to have been attacked by cookiecutters is long. It includes marlin, broadbill swordfish, tunas, opah, pomfrets, some sharks, many species of whales and dolphins, wahoo, dolphinfish, oarfish and probably many others.

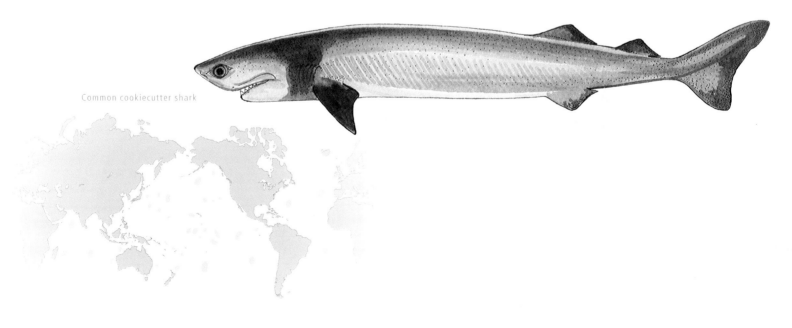

Common cookiecutter shark

The telltale
teeth of the
cookiecutter
shark, razor-sharp
and able to take
out a plug of flesh
in an instant
*seapics.com/
Gwen Lowe*

The gruesome-
looking crater
wound left by
a cookiecutter
shark on a small
swordfish. These
wounds are found
on many species
of open-ocean
fish.
John Diplock

WHALE SHARK
Rhincodon typus

OTHER COMMON NAMES

France	Requin baleine
Japan	Jinbeizame
Spain	Tiburón ballena

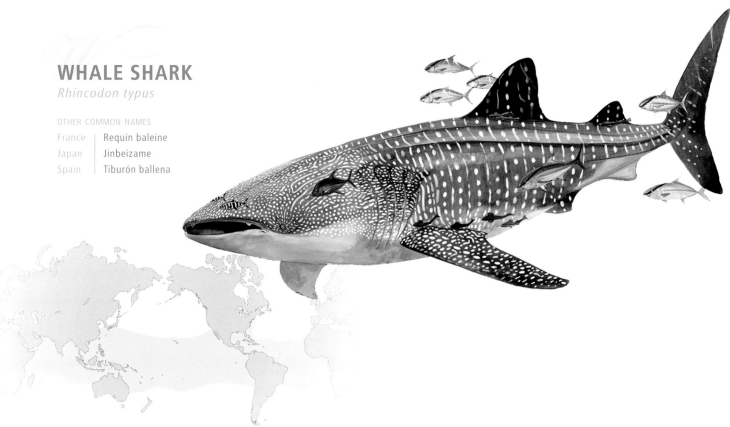

The whale shark is the largest living fish, and one of three large filter-feeding sharks. The other two are the basking shark, also a creature of the surface, and the little-known megamouth shark, which dwells in deeper water. The relationships of the whale shark were clouded for some time, but recent work suggests that it should be placed in the Orectolobiformes, a group which includes the carpet sharks and zebra shark (*Stegastoma fasciatum*). The latter bears some resemblances to the whale shark in its body markings and presence of longitudinal body ridges.

The whale shark is unmistakable. Apart from its large size, it has an extremely broad, flat head, a wide, nearly terminal mouth, several prominent longitudinal ridges on the back and is marked with characteristic whitish spots and bands in the form of a 'checkerboard' across a blue-gray back. Whale sharks have thousands of tiny teeth, less than 6 mm long, arranged in up to 300 rows in each jaw.

This is a cosmopolitan (worldwide) species with a tropical/warm temperate distribution extending about 25° north and south of the equator. Whale sharks prefer surface temperatures above 22°C, and are regular visitors to tropical coasts,

even entering coral lagoons. They are especially associated with upwellings where tropical krill, other crustaceans and small fishes such as deep-water lanternfish aggregate in the nutrient-rich water. In fact, seasonal aggregations of whale sharks are highly predictable, based on such concentrated food sources.

Movements of whale sharks are poorly understood, although some satellite tagging has shown that they can move over thousands of kilometres, resulting in the species being classified as highly migratory. Tagging with conventional tags has shown some short-term site fidelity, and also a tendency for at least some animals to return annually to the same areas.

Knowledge of whale shark movements and stock structure in the Indian Ocean is particularly important since they are fished commercially off India and in the Philippines. At Ningaloo Marine Park, on the coast of northwestern Australia, an annual aggregation of perhaps hundreds of whale sharks is a big tourist drawcard, and also presents the opportunity for scientists to attach archival tags to them. Annual returns of conventionally tagged fish suggest that these data-storing tags might be able to

The awesome sight
of a whale shark,
the largest fish in
the ocean, filter-
feeding as it swims
slowly near the
surface
James Watt/
imagequestmarine.com

be retrieved by divers a year later, thereby providing detailed information on their movements in the intervening period. Whale sharks also appear with considerable regularity in the Coral Sea off the Great Barrier Reef. There, they are often associated with seasonal aggregations of lanternfishes on which they feed, and which also attract large numbers of bigeye and yellowfin tuna. In this area, whale sharks seem to act as natural fish aggregating devices and it is not uncommon to see large schools of bigeye tuna swimming in a vertical column beneath each whale shark.

Although growing to a reputed 18 metres, the generally accepted maximum measured length of a whale shark is actually 13.7 metres, with the most recent accurate maximum measurement being 12.1 metres. The weight of such huge fishes is very difficult to measure, but there is no shortage of estimates of maximum weights in the literature, ranging from 20 to 40 tonnes. Size at first maturity has been estimated at about 8 metres. No estimates of growth rate are available, but it is assumed that growth is very rapid in early years.

For many years, whale sharks were thought to be oviparous, meaning that they lay eggs which would subsequently hatch on the sea floor. This was based on two lines of thought. Because some of their relatives, such as the carpet sharks and leopard sharks, lay eggs, it was thought reasonable to assume that the biggest member of the clan might do the same. The second piece of evidence was a single whale shark egg which was trawled in the Gulf of Mexico in 1953. This egg case (often referred to as the largest egg of any animal) contained a single embryo measuring 36 cm in length. However, the papery thin egg case and lack of tendrils which shark eggs usually possess to anchor to the substrate led some scientists to suspect that this may have been an aborted egg. The true mode of reproduction of the whale shark was finally revealed in 1995 when a large female specimen (about 11 metres long) was dissected at a Taiwanese fish market and found to contain 300 embryos ranging from 42 to 63 cm in length. This is by far the largest litter size for any shark species. These embryos had, and were hatching from intrauterine eggs, thus proving once and for all that whale sharks are indeed ovoviviparous.

The whale shark is a harmless filter feeder, although it does not necessarily passively strain its food by moving slowly through the water like some of the baleen whales and the other filter-feeding shark, the basking shark. Rather, it is primarily a suction feeder, often feeding in a vertical position with its head towards the surface, suddenly sucking in large volumes of water containing its food items. Usually its food consists of small fishes and crustaceans, but they have also been recorded with larger fish such as tuna and remoras in their stomachs – perhaps inhaled accidentally rather than by design.

Whale sharks have been, and are still hunted for food in some areas. In one fishery off India, it is estimated that as many as 1000 were taken in 1999–2000, although this is difficult to verify. In the Philippines, some villages specialize in hunting whale sharks for local consumption, as well as for export to Taiwan. The numbers taken throughout the Philippines are not known, but the annual harvest would almost certainly number in the hundreds. Following pressure from conservation groups, the killing of whale sharks (and manta rays) was banned by the Philippine Government in 1998. However, the effectiveness of the ban is questionable, since whale shark meat destined for export is still regularly confiscated at airports in the Philippines. The global number of whale sharks appears to be declining, to the extent that the species has been listed as vulnerable on the International Union for the Conservation of Nature red list.

BASKING SHARK
Cetorhinus maximus

The basking shark, a slow-swimming planktivorous shark, is the second largest fish in the ocean, growing to lengths of at least 9 metres, possibly larger. The maximum size of the basking shark is only surpassed by another filter-feeding shark, the whale shark, *Rhincodon typus*.

Free-swimming basking sharks may be superficially confused with great white sharks, *Carcharadon carcharias*. Both are members of the same order, Lamniformes, and share broad lateral caudal keels, a nearly symmetrical crescent-shaped tail and a generally similar body shape. However, apart from its usually huge size, the basking shark is readily identified by its extremely long gill slits, extending nearly around the body, its elongated cone-shaped snout and its vast underslung mouth armed with myriads of tiny hooked teeth.

The basking shark has a worldwide distribution, tending to prefer waters over continental shelves and slopes, but also occurring around major island groups such as Hawaii, the Faeroes and the Galapagos. While usually associating with land masses, it is also encountered in mid-ocean locales and satellite tracking studies have confirmed such offshore movements. Its main concentrations are in the cooler subtemperate or boreal waters of both the northern and southern hemispheres, with preferred temperatures of 8°C to 14°C, although it is known to acclimate to temperatures up to 26°C in some areas.

After the whale shark, the basking shark is clearly the second largest fish on the planet. However, as is the case with other large sharks, determining the largest verified size of the basking shark is problematic. Sizes up to 12, or even 15 metres are cited, but since these are not proven, it is somewhat surprising that they continue to be repeated. In fact, aerial estimates of thousands of basking sharks indicate that individuals over 8 metres long are uncommon, and that more realistic maximum lengths are about 9 metres for males and 9.8 metres for females. A maximum weight of 4000 kg (8800 lb) is also often cited for the species, and while this seems quite feasible, again it is unverifiable.

Very little is known about basking shark reproduction. Very few pregnant females have been observed (one contained six young) and it is thought that the mode of reproduction is likely to involve intrauterine 'oophagy', whereby growing young eat eggs inside the uterus

which are continually produced by the mother. Gestation may be as long as 3.5 years, and size at birth is thought to be 1.5 to 1.7 metres long.

In Europe, it was historically postulated that basking sharks shed their gill rakers in early winter and then hibernated near the bottom for the winter and early spring before reappearing on the surface. This theory, however, was at odds with long-term aerial observations off southern California where basking shark sightings were common over winter and spring.

As noted, basking sharks feed exclusively on zooplankton. Satellite tagging over the past decade or so has shown that sharks do not hibernate, but undertake extensive movements of thousands of kilometres, as well as regularly diving to considerable depths (greater than 750 metres). Furthermore, basking sharks have been shown to locate feeding grounds by active rather than passive foraging. Comparisons of actual movements of basking sharks with computer simula-

tions of random foraging indicate that the sharks are very good at finding zooplankton communities both at the surface and in midwater. It is also suggested that this skill in finding food may be learned rather than innate.

In the past, and in some areas today, basking sharks have been and are hunted for meat (dried and salted), fishmeal, fins, liver (for squalene) and their skin (for leather). One rather bizarre 'fishery' for the species occurred off British Columbia in the 1950s when the Canadian Government commissioned a boat with a sharp plough-like construction on the bow to ram and kill basking sharks because they were damaging gill nets set for salmon. It was reported in the first month of operation that 50 sharks were killed in this manner.

There have been concerns over possible declines in numbers of basking sharks in recent years, and the species is now listed as vulnerable by the International Union for the Conservation of Nature.

PELAGIC STINGRAY
Pteroplatytrygon violacea

The pelagic stingray is the only ray to have evolved a truly pelagic lifestyle, and in doing so, has successfully spread to occupy a very broad global range.

This is a distinctive ray best identified by the dark purple coloration on its upper surface, and lighter purple on the underside. In other characteristics, though, it resembles its close relatives in the stingray family Dasyatidae, including the possession of a long venomous spine a third of the way along the whip-like tail. It has a broad, rounded snout and eyes that are flush with the body contours.

The pelagic ray is a cosmopolitan species, present in tropical to subtemperate waters of all oceans and the Mediterranean Sea.

Growth rates of the pelagic ray in the wild have not been studied, but observations of a number of individuals held in captivity at the Monterey Bay Aquarium have given some insights. Rays weighing about 2 kg (4.5 lb) and measuring 40 to 45 cm were estimated to be about two years old. In this study, the largest male attained a disc width of 68 cm and weight of 12 kg (26 lb) at an estimated seven years of age, while the largest female, which weighed 49 kg (108 lb) and measured 96 cm across the disc, was estimated at nine years old. This latter size appears to be the largest yet recorded for the species.

Pelagic rays are born live after hatching from parchment-like eggs inside the mother. Litter sizes range between two and nine, and full-term young vary between 200 and 400 grams in weight. Observations in captivity suggest it is likely that pelagic rays give birth to two litters per year.

Because it lives most of its life in the upper 100 metres of the open ocean, the pelagic ray is a common bycatch of longline and driftnet fisheries. The actual catches and impacts on populations, however, are entirely unknown.

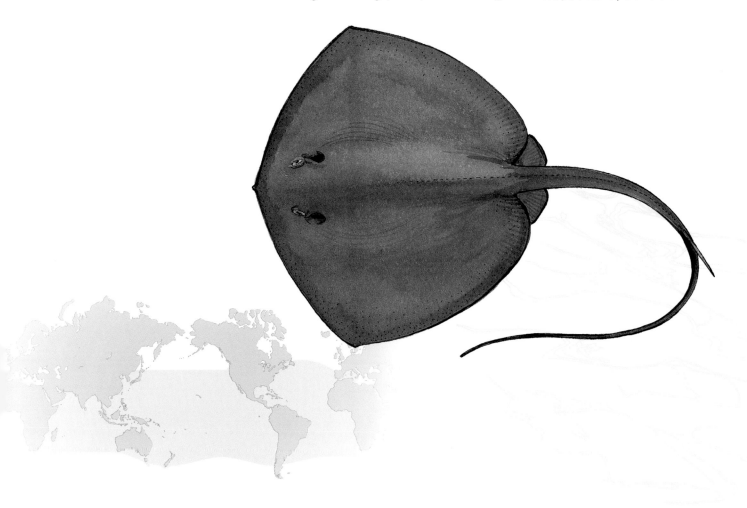

DEVILRAYS
Manta birostris,
Mobula spp.

OTHER COMMON NAMES

	Manta ray
	Devilfish
France	Mante géante
Japan	Oni-itomaki-ei
Spain	Manta voladora

The devilrays (family Mobulidae) are large, pelagic rays, one of which, the manta ray, is by far the largest ray in the world. Until recently, ten species of devilray were recognized around the world – the manta ray, *Manta birostris*, and nine species in the genus *Mobula*. However, another species of giant manta, possibly two, have only recently been discovered. The new species is widespread but has not yet been described or named, so this account deals primarily with *Manta birostris*.

The devilrays are diamond-shaped with very pointed pectoral tips, and possessing a pair of characteristic extensions of the pectoral fins (called cephalic lobes) extending forward from either side of the head in front of the eyes. These rays possess a short whip tail, some species with a serrated sting (although not the giant manta ray). Devilrays also possess a very small dorsal fin above the origin of the whip tail.

The members of the devilray group have tropical to temperate distributions, being found mostly over continental shelves and around larger island groups. The manta ray is widely distributed, extending along continental coasts to about 35°, but mainly confined to tropical and warm temperate regions. These rays are creatures of the surface, only diving to shallow depths of 30 metres or so.

Little is known about the movements of the manta ray. It can certainly swim at high speeds, and has the appearance and swimming stroke capable of steadily covering long distances.

The giant manta ray is one of the largest fish that swims, although just what size it might actually attain is difficult to ascertain. Specimens certainly reach widths of at least 6.7 metres, and a measurement of 9.1 metres is reported in the literature. However, as with many reputed lengths of big fish, this latter figure converts to an exact round number (30 feet), and is therefore likely to be an estimate rather than an actual measure. As noted below, a female manta ray caught in the Galapagos was apparently reliably measured at 5.5 metres and weighed 1050 kg (2310 lb). A maximum recorded weight of 1400 kg (3080 lb) is mentioned in some accounts, but again, this is a neat, round number and, without verification, is also likely to be an estimate.

Manta rays are ovoviparous, meaning that their young develop in eggs inside the mother, and hatch some time before being born alive. They have very

Giant manta ray

low rates of reproduction, with a gestation period of about 13 months, only giving birth to one or two young. The young are large at birth, measuring about 1.5 metres between the tips of the pectoral fins. The large 5.5 metre female mentioned above was found to contain a single embryo weighing 12 kg (26 lb).

Like some other huge marine animals, such as baleen whales, whale sharks and basking sharks, manta rays are filter feeders. It is when they are feeding that the function of the cephalic lobes becomes clear. These are folded downwards to form a funnel in front of the mouth, which opens wide as the ray swims upwards in a spiral, filtering plankton through its sponge-like gill plates. It has been estimated that manta rays consume about 2 per cent of their body weight in plankton each day, which, for a 1000 kg (2200 lb) animal, is about 20 kg (44 lb) per day.

Manta rays often form 'squadrons' of associated individuals which travel in formation, sometimes spread over quite

large areas. When swimming near the surface, the tips of their pectoral fins often break the water, giving the appearance of the dorsal fins of sharks. Manta rays have often been observed leaping clear of the water. The reason for this behavior is unknown, although it has been suggested that it may be related to courtship.

There are minor artisanal fisheries in some Indo-Pacific areas (primarily in Mexico and the Philippines) that target manta rays by harpooning from open boats. Manta rays are also occasionally taken as a bycatch of purse seine and longline fisheries, although the numbers caught are believed to be small. There are also reports of manta rays taking live fish baits in recreational fisheries, although again, such cases are rare. The status of populations of the manta ray is uncertain, although with such a low population turnover rate, there are some concerns about continuing exploitation. The recent discovery of at least one other species of giant manta further complicates considerations of population status.

SPOTTED EAGLE RAY
Aetobatus narinari

There are many species of rays around the world but only relatively few that are highly active swimmers and which spend at least part of their time at the surface, thereby periodically entering the pelagic habitat. The eagle ray is one such species, often observed swimming rapidly at the surface, at times even leaping clear of the water.

The eagle rays, and their close relatives, the cownose rays, are highly distinctive. Unlike other rays, they have a distinct head which protrudes in front of their diamond-shaped bodies. The cownose rays have a flattened snout with a marked indentation in the middle, while the eagle rays have a rounded snout. The spotted eagle ray is further identified by its prominent fleshy, upturned snout and its distinctive small white spots all over the dorsal surface. It has an extremely long whip-like tail with a strong stinging barb near its base. There are four species of eagle ray, but the spotted eagle ray is by far the most widespread.

The spotted eagle ray has a truly global distribution in tropical to temperate waters. It tends to be a coastal species, preferring clear, marine waters, especially around reefs and atolls. Because of its active swimming habit and its broad distribution which includes offshore archipelagos, it is thought likely to be capable of crossing ocean basins. Some authors consider that there may be more than one species of spotted eagle ray, but this is yet to be proven.

This ray grows to a large size. The maximum 'disc width', the distance between the two pointed pectoral flaps, is often cited as 330 cm, and the maximum length, from the tip of the snout to the tip of the whip-like tail, as 880 cm.

Like many other rays, the spotted eagle ray gives birth to live young. Litter sizes are between two and four and the young are born with a disc width of about 25 cm.

As is the case with many rays, the spotted eagle ray tends to be caught as a bycatch in other fisheries. The main method of capture would be coastal gill nets, although it is also taken on longline. The species is one of the relatively few rays which are sought by sport fishers.

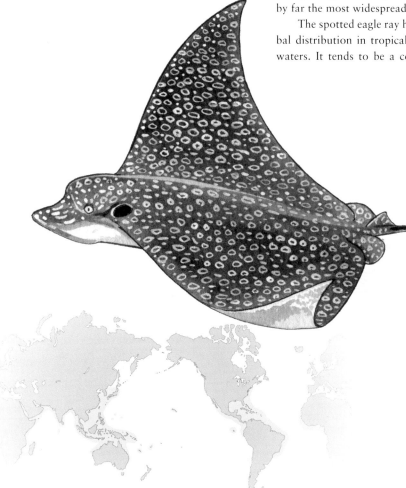

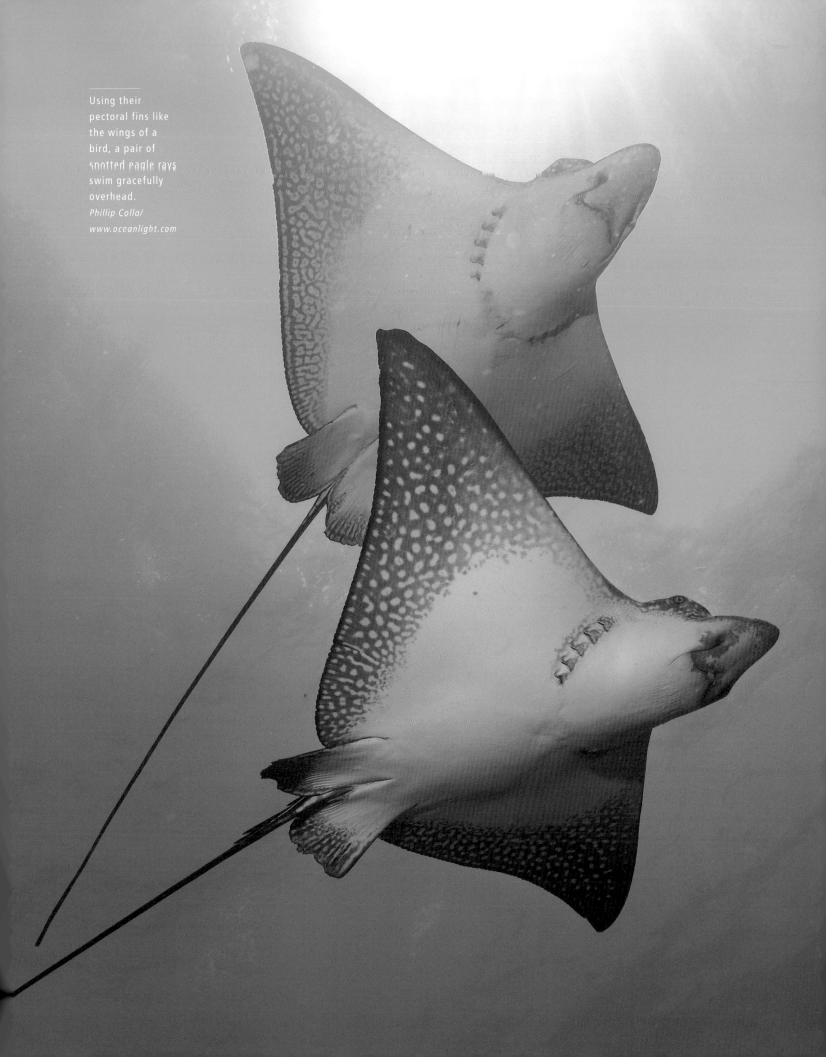

Using their pectoral fins like the wings of a bird, a pair of spotted eagle rays swim gracefully overhead.
*Phillip Colla/
www.oceanlight.com*

11

OTHER OPEN-OCEAN FISHES

Other open-ocean fishes

REMORAS
Family Echeneidae

OTHER COMMON NAMES
Suckerfish

Common remora *Remora remora*
France and Spain | Rémora
Japan | Nagakoban

White sucker fish *Remorina albescens*

Live sucker fish *E. naucrates*

The remoras are among the most unusual fishes of the open ocean. The eight known species are all closely related members of the family Echeneidae, all of which possess a remarkable sucking disc on top of the head. This elaborate structure is, in fact, a highly modified first dorsal fin. The sucker has a series of transverse plates that are modified fin spines. And like the movable spines of a normal dorsal fin, the plates can be raised and lowered at will. When a remora wants to attach itself to any object, it makes firm contact by means of the fleshy circumference of the disc and then raises the plates slightly, effectively creating a series of vacuum chambers. The resulting adhesion is so strong that it is almost impossible to prize the fish off – the only method seeming to be to slide the fish forward until the suction is broken. The attachment needs to be firm, of course, since the remoras spend most of their time firmly stuck to the sides of 'host' species of large fish, cetaceans or sea turtles. In so doing, they obtain a free ride to new areas, gain protection from predators, and more importantly, feed on the hosts' parasites, as well as leftover food particles.

Most of the remora species are specialized with respect to the host animals they choose to cling to. Thus, there is the whalesucker, *Remora australis*, which only attaches to dolphins and whales, the spearfish remora, *Remora brachyptera*, which attaches itself to the inside gill plate of marlin and sailfish, the white suckerfish, *Remorina albescens*, which is almost exclusively found on manta rays, and the slender suckerfish, *Phtheirichthys lineatus*, which is most often found attached to barracudas. On the other hand, the two species of remora illustrated here tend to be exceptions to this rule in that they are observed on a range of hosts. The common remora, *Remora remora*, the largest of the group, is often seen on various shark species, including the whale shark, and on large manta rays. On the latter, a pair of large common remoras often take up stations symmetrically on either side of the manta's disc, presumably helping to balance the swimming 'flight' of their host. Lastly, the live sharksucker, *Echeneis naucrates*, is found not only on sharks but also bony fishes,

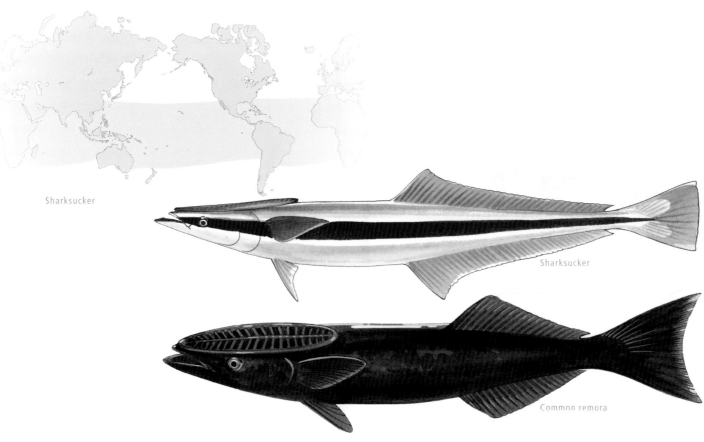

Sharksucker

Sharksucker

Common remora

whales, dolphins and even ships. Recent observations in the wild have reported juvenile sharksuckers attaching to many fish species, including porcupine fish, snappers and larger members of their own species.

A fish that bears a striking resemblance to one species of remora, the live sharksucker, is the cobia, *Rachycentron canadum*, itself a worldwide species and treated separately in this book. Both have very similar body shapes, including a broad head, similar fin placement, tail shape, and even striped coloration. Neither the remora family nor the single species cobia family have any obvious, close relatives, but many scientists suspect that the remora and the cobia may well have evolved from a common ancestor. In fact, cobia have a habit of swimming very close to sharks and manta rays, so it is not a huge leap of speculation to imagine the link becoming much closer over evolutionary timescales. On the other hand, it is possible that cobia have evolved as mimics of the live sharksucker, so that it is afforded the same protection and advantages of the latter in associating with large fish. No doubt, genetic studies will resolve this interesting problem in the long run.

Remoras have been known and recorded since ancient times. The Latin word 'remora' means 'a hindrance or delay', and the Greeks and Romans certainly believed that remoras could impede the smooth progress of sailing ships, or even stop them altogether. The Roman author Pliny wrote 'why should our fleets and armadas at sea make such turrets on the walls and forecastles, when one little fish is able to arrest and stay our goodly and tall ships?'

Top The common remora (top) and the spearfish remora (bottom), showing the remarkable sucking disc, actually a highly modified dorsal fin
Julian Pepperell

Lower Lemon sharks with large numbers of attendant remoras, in this case, live sharksuckers
Guy Harvey

FLYING FISHES
Family Exocoetidae

OTHER COMMON NAMES

France | Exocet
Japan | Tobi
Spain | Pez volador

There are well over 50 species of flying fishes worldwide, perhaps as many as 67, all of which belong to the single family Exocoetidae. All are torpedo-shaped, silvery fish, and it is generally believed that they evolved from a common ancestor of the halfbeak family – the so-called halfbeaks or ballyhoo, Hemiramphidae, which they resemble. There are, in fact, several species of flying halfbeaks, the most common in the Pacific being the long finned garfish (also called the ribbon halfbeak), while a closely related species in the Atlantic Ocean, the flying halfbeak, has somewhat larger pectoral fins. Flying fishes are members of the Beloniform group of fishes which, as well as halfbeaks, also includes the needlefishes and the sauries.

The flying fishes are divided into two types: Four-winged and two-winged flying fishes. The two-winged forms belong to the genera *Exocoetus* and *Parexocoetus*, while the other genera are all four-winged. The huge lateral 'wings' of flying fishes are highly modified pectoral fins, but in four-winged flying fishes, the paired pelvic fins, further back along the underside of the body, are also greatly enlarged and fully extend during flight. Depending on the species, adult flying fish can range in size from 15 to 50 cm in length.

As a group, the flying fishes are distributed throughout the tropical and subtropical to temperate oceans of the world. All are pelagic species, some with relatively coastal and quite restricted distributions while others are completely oceanic and very widespread. Examples of the latter would include the barbel flying fish, *Exocoetus monochirrus* (a two-winged species), and the black wing flying fish, *Hirundichthys rondeletii* (a four-winged species).

For a fish which spends most of its life in surface marine waters, it is somewhat surprising to learn that the eggs of most species of flying fish are heavier than water. This means that the only way flying fish spawn successfully is by attaching their eggs to floating objects, especially seaweed fronds such as kelp. This is achieved by means of long, sticky filaments on the eggs. Using this knowledge, the Japanese have developed a strategy of harvesting flying fish eggs from the wild by using 'artificial kelp' – to ultimately appear on sushi plates as glistening (and delicious) orange flying fish roe.

The flight of a flying fish is a wondrous thing. Beating the tail at a blurring rate of 50 to 70 beats per second, flying fish achieve take-off speed at about

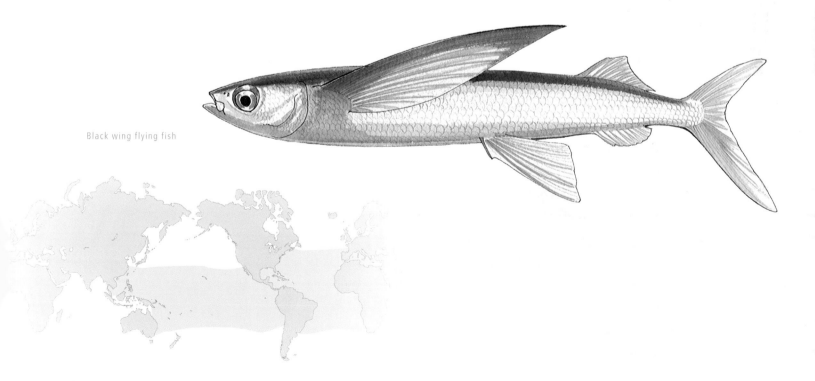

Black wing flying fish

THE FLYING FISH.——*Exocetus volitans.*

35 km/h, and once airborne, reach air-speeds of up to 70 km/h. For many years, it was speculated as to whether flying fish flapped their 'wings' or not. This was finally resolved when film of flying fish shot at night under strobe lighting clearly showed that the pectoral fins do not flap, but are used only as gliding airfoils.

The longest flights recorded are up to 400 metres, although such distances are achieved by dipping the long lower lobe of the whirring tail back into the water a couple of times during the flight for extra bursts of propulsion. It is also thought that the 'ground effect' – that is, pressure between the wings and the ocean surface – may help keep the fish airborne, as is the case for the effortless ocean-skimming of many species of seabirds. While flights of flying fishes are usually 1 or 2 metres above the surface, there are many obser-vations of fish landing on the decks of boats up to 13 metres high.

Flying fish only occur in warm waters (above about 20°C), and it has been cal-culated that they are probably unable to summon the energy required to achieve flight at lower temperatures. Scientists are still uncertain whether flying fish have control over the direction of their flight or not. They don't seem to maneuver much in midair, but their eyes look downwards and can almost certainly focus in both air and water. This is also borne out by the fact that they often seem to intentionally land in weed lines (good sources of food), so it is likely that they can control their flight to some extent. On the other hand, it is not uncommon for flying fish to land on the decks of boats in broad daylight, so flight control may not include the abil-ity to avoid large floating objects.

An early 19th century illustration of a two-winged flying fish – not anatomically correct, but a delightul depiction nonetheless

A stunning photo
of a flying fish
gliding towards a
touchdown on a
glassy sea
Geoff Jones

SAURIES
Scomberesox spp.,
Cololabis spp.

OTHER COMMON NAMES
France	Balaou
Japan	Sanma
	Saira
Spain	Paparda

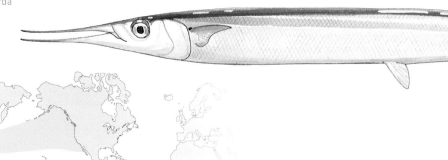
Pacific saury

Pacific saury

Atlantic saury

Southern saury

There are just five species of saury in the world. They are the Atlantic saury, *Scomberesox saurus saurus*, the southern saury (or king gar), *Scomberesox saurus scombroides*, the dwarf saury, *Scomberesox simulans*, the northern Pacific saury, *Cololabis saira*, and for completeness, the rather obscure eastern Pacific saury, *Cololabis adocetus*. Often linked with the halfbeaks or garfishes, it has recently been shown by genetic studies that the sauries are in fact more closely related to the needlefishes, or longtoms (family Belonidae).

All the sauries are beautiful, intensely silver fish with dark blue backs – the classic pelagic countershading indicating their existence in the surface waters of the open ocean. They have powerful stiff tails and a series of finlets behind their dorsal and anal fins, just like mini tunas (in fact, the generic name *Scomberesox* can be roughly translated as 'tuna pike').

Both the Atlantic and the southern saury have two thin toothless beaks or bills – in contrast to the northern Pacific saury, which lacks these beaks.

The southern saury is the most widespread of the group, occurring right around the southern hemisphere, mainly between 30°S and 40°S. As their names suggest, the Atlantic saury and the northern Pacific saury are confined to those ocean areas.

Studies on the northern Pacific saury, and which probably apply to other species, have shown them to be short-lived, with the largest fish (about 40 cm long) possibly being as young as 14 months old (although other studies suggest a maximum age of up to four years).

Sauries, especially the southern species, are not caught in large numbers mainly because large schools are sporadic and unpredictable. Water temperatures and climate events such as El Niño have been shown to have marked effects on the migration and growth rates of sauries, again mitigating against the establishment of regular commercial fisheries. Sauries are rarely caught by recreational anglers, primarily because they are planktonic feeders, eating zooplankton (mainly tiny crustaceans and snails), fish larvae and fish eggs.

Sauries are very important components of the diets of not only tuna, billfish and oceanic sharks, but also marine mammals such as dolphins and fur seals and seabirds, including penguins, terns and gannets. In one recent study, cape gannets off South Africa were estimated to consume over 6000 tonnes of sauries per annum. One can only imagine what the total consumption of sauries by all their predators must be.

One of the most important of the world's so-called baitfishes, the sauries (in this case, a southern saury) are open-ocean fishes that form a major component of the diet of many predatory fishes.

NEEDLEFISHES
Family Belonidae

The needlefishes (also called longtoms in some regions) are members of the Beloniform group of fishes which also includes the halfbeaks, sauries and flying fishes. There are at least 34 species of needlefishes globally, ranging in habitat from freshwater to oceanic. The three species chosen here are typical of the more offshore species.

Both jaws of the needlefishes are very elongated, and armed with sharp teeth used to seize prey fish. Needlefish are very surface-oriented, lying in wait for prey just below the surface. They hover and then strike quickly with a sudden forward burst of speed.

Hound needlefish
Tylosurus crocodilus crocodilus

This is one of the most widespread of the needlefishes, occurring in all three major oceans – often well offshore, including around remote island groups. This species is one of the largest of the group, growing to at least 1.5 metres in length, sometimes cited up to 2 metres. It has a relatively stout, cylindrical body and a shorter head compared with other needlefishes.

This species has been implicated in injuries and even deaths caused by startled fish leaping from the surface at night and accidentally impaling people traveling in open boats. As with other needlefishes, the hound needlefish is quite edible, but their greenish colored bones tend to dissuade fussy eaters.

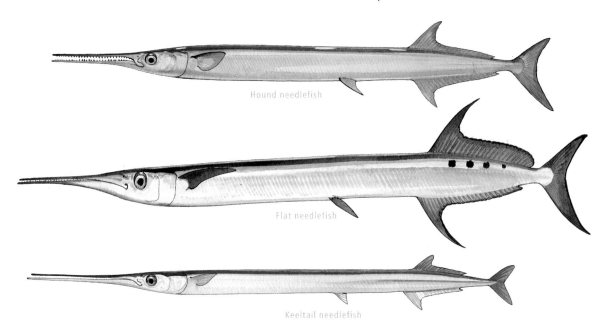

Hound needlefish

Flat needlefish

Keeltail needlefish

Hound needlefish

Flat needlefish
Ablennes hians

The flat needlefish is also common in all three major oceans, but is not as oceanic in its distribution as the hound needlefish. It is readily identified by its laterally compressed, body, very tall, curved dorsal and anal fins and dark vertical barring on the upper flanks, especially towards the rear of the body.

This is also a large needlefish, growing to about 1.4 metres in length. The IGFA all-tackle angling record is a fish of 4.8 kg (10.6 lb) caught off Mozambique in 1997.

Keeltail needlefish
Platybelone argalus platyura

The keeltail needlefish is also widespread, but with a patchy distribution extending to the open ocean, including remote islands such as the Hawaiian archipelago. It is common in reef lagoons and over pristine sand flats. It is most easily identified by the presence of a single large lateral keel on either side of the caudal peduncle (tail wrist).

This is quite a small needlefish, only growing to about 40 cm long.

HALFBEAKS
Family Hemiramphidae

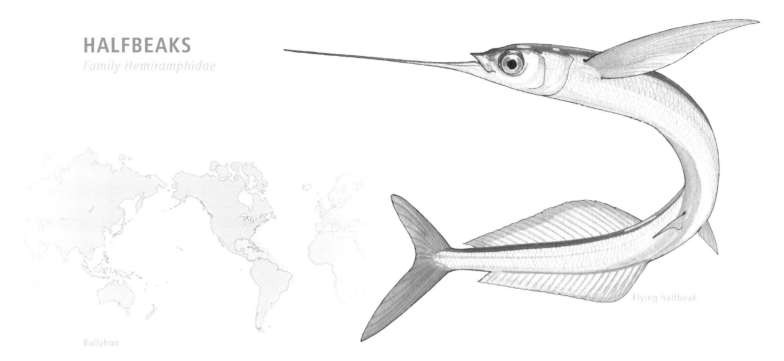

Ballyhoo

Flying halfbeak

The halfbeaks, also known as garfishes in some parts of the world, are members of the Beloniform group which also includes the flying fishes (Exocoetidae), needlefishes (family Belonidae) and sauries (Scomberesocidae). There are at least 85 species of halfbeaks globally, ranging in habitat from freshwater to oceanic. Nearly all the halfbeaks have a very short upper jaw and usually an extended lower jaw. The first two outlined here are typical of the more inshore species, commonly used as trolled baits, while the third is one of the oceanic flying halfbeaks.

Ballyhoo
Hemiramphus brasiliensis
This well-known species occurs extensively in both the western and eastern Atlantic, mainly inshore, but is also found in large schools further offshore. As with a number of species of halfbeaks all over the world, these are a favorite bait for offshore sport fishers targeting billfishes, especially sailfish and marlin. Like many other halfbeaks, the ballyhoo is at least partially vegetarian, feeding on floating seagrass as well as small crustaceans and fish. It is reported to grow to 55 cm overall length, but this seems excessive. Most would be around 30 cm long.

Balao halfbeak
Hemiramphus balao
The balao has a similar distribution to the ballyhoo. The two species can be separated by the length of the pectoral fin. In the ballyhoo, the pectoral fin, when folded forward, reaches the nasal pit; in the balao, it does not. In addition, the ballyhoo has a very short upper jaw and a red tip to its beak. It attains about 40 cm in total length.

Flying halfbeak
Euleptorhamphus velox
The two species of flying or longfinned halfbeaks are characterized by their very long lower jaws and long pectoral fins. These remarkable halfbeaks skitter along the surface, with pectoral fins extended to obtain lift. Then, once clear of the water, rather than use their pectoral fins to glide, they bend their bodies into a crescent, forming an airfoil of sorts which allows them to glide into the wind, using their narrow bodies for lift.

The ribbon halfbeak (also called the longfinned or flying garfish), *Euleptorhamphus viridis*, is a worldwide pelagic species, while the closely related flying halfbeak, *Euleptorhamphus velox*, is confined to the Atlantic.

The ribbon halfbeak grows to 61 cm, while the flying halfbeak attains 53 cm.

PACIFIC SARDINE
Sardinops sagax

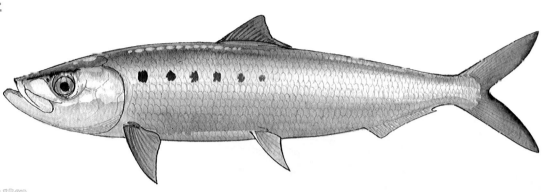

The Pacific sardine, a member of the herring family, Clupeidae, is an extremely important forage or baitfish in near oceanic and inshore waters of the major continents in both northern and southern hemispheres. It is absent from the Atlantic, except for its occurrence along the southwestern African coast. It is a filter feeder, eating mostly planktonic crustaceans, but also phytoplankton including diatoms.

There are regional differences in morphology and maximum sizes of this species which almost certainly reflect discrete stocks. Even though subspecies status has been suggested for some of these regional populations, the Pacific sardine is still regarded by most authorities as a single species throughout its range.

Quite extensive coastal migrations have been shown for the species in some areas, notably along the Californian coast. The Pacific sardine is thought to be remarkably long-lived for such a small species, perhaps attaining ages of up to 25 years.

Historically, catches of Pacific sardines have been very large, but have fluctuated greatly, and while overfishing may have occurred in some locations at some times, it is thought that environmental influences such as the El Niño/La Niña cycle play a major role in influencing its distribution and abundance.

ATLANTIC HERRING
Clupea harengus harengus

OTHER COMMON NAMES

	Yawling
France	Hareng de l'Atlantique
Spain	Arenque del Atlántico

Like the Pacific sardine, another clupeid, the Atlantic herring plays a vital role in the pelagic food chain of the northern Atlantic – being, for example, a favored food item of Atlantic bluefin tuna. It occurs near the coast, but also in continental shelf waters of northeast United States, Canada, Greenland and northern Europe.

Also in common with the Pacific sardine, there are very likely separate stocks or even races of Atlantic herring, even within the relatively small geographic regions.

Atlantic herring feed mainly on zooplankton, but are also capable of filter-feeding on smaller phytoplankton. Atlantic herring do not produce planktonic eggs as sardines do, but rather, lay them in masses on substrate in coastal waters.

Like the sardine, the herring is quite long-lived for a forage fish, with estimates of age at maturity ranging from three to nine years. This is a very prolific fish which nevertheless has been historically overfished in some areas. Happily, though, the stocks of herring along the Atlantic US states are now considered to be in good shape, with more than 80 000 tonnes landed in 2007.

BLUEBOTTLE FISH
Nomeus gronovii

OTHER COMMON NAMES

	Man-of-war fish
	Shepherd fish
Japan	Eboshidai
Spain	Pastorcillo

The bluebottle fish (also known as the man-of-war fish) is a member of the family Nomeidae, collectively known as driftfishes. There are 18 species in the family, but the bluebottle fish is probably the most common and widespread. Most of the driftfishes associate with floating material and debris, whereas the bluebottle fish is nearly always found in the company of oceanic jellyfishes, in particular, the drifting colonial jellyfish (siphonophore), the so-called Portuguese man-of-war, or 'bluebottle' (*Physalia physalis*). *Physalia* is colored with intense blue and indigo pigments, and the bluebottle fish has similar coloration, mottled with dark blue to violet patterns which obviously provide camouflage among their hosts' tentacles. Apart from their color, bluebottle fish have greatly enlarged fan-like pelvic fins which are attached to the belly by a membrane along their entire length.

The bluebottle fish apparently eats the tentacles of its protective bluebottle hosts, but also feeds on invertebrates which also associate with the floating jellyfishes. While the bluebottle fish clearly obtains shelter, protection and food from its invertebrate hosts, it seems that there is no obvious benefit which might be provided by the fish to the jellyfish, other than possibly defending them from predation from other fishes or perhaps by attracting other fish for the jellyfish to feed on. However, this has not been proven, suggesting that the relationship may simply be one-sided.

While every depiction of this species in the literature shows the mottled blue form, it is believed that this, and other pelagic driftfishes, may simply be juvenile phases which eventually settle to the bottom as adults. In the case of the bluebottle fish, so rare are presumed adults that a Japanese paper published in 1986 described just one possible adult specimen. The 24.1 cm fish was caught on the bottom by trawling. It was not blue, but a dull brown, and only identified as a bluebottle fish because of its matching vertebral and fin ray count (it did not have an expanded pelvic fin, which was assumed to be due to differential growth). Interestingly, the maximum size of the bluebottle fish most often quoted in the literature is 39 cm; however, the original source of this figure does not provide a citation or any other information. The true nature of the life cycle of this delightful little fish therefore still remains somewhat of a mystery.

LANCETFISH

Alepisaurus ferox,
A. brevirostris

OTHER COMMON NAMES

	Handsaw fish
	Wolf fish
France	Cavalo ocelle
Japan	Mizu-uo
Spain	Lanzón picudo

Resembling sailfish with fangs, the two species of lancetfish, the only members of the family Alepisauridae, are common denizens of the mesopelagic (midwater) zone of the open ocean, and are among the most common bycatch species of the world's longline fisheries. However, because the flesh is soft and spongy, virtually all lancetfish are discarded at sea and so are rarely seen by the general public.

These are unmistakable fishes with very high, long-based first dorsal fins followed by a small adipose (fatty) fin towards the tail wrist. The teeth are distinctive, with several long semi-transparent fangs in each jaw. There is a delicate whip-like extension of the upper lobe of the tail, although this is often broken off during capture. As the names suggest, the longnose lancetfish (*Alepisaurus ferox*) has a relatively long and pointed snout compared with the shortnose (*Alepisaurus brevirostris*). Both the longnose and the shortnose lancetfish have extremely broad global distributions, extending from high latitudes in both hemispheres right through to the tropics.

Very little is known about the biology of the lancetfishes. They grow to relatively large sizes, the largest shortnose attaining 96 cm in length and the long-

nose, 215 cm. The longnose lancetfish is known to occur from the surface (at night) to as deep as 1800 metres. That they do spend time at depth is indicated by the fact that, like many deep-water fish species, the swim bladder of the lancetfishes is filled with oil rather than gas. The oil provides the fish with buoyancy at depths where gases would be compressed to very small volumes.

Examination of the lens and microstructure of the retina of lancetfish suggests they have excellent eyesight. They are certainly voracious and it is not uncommon on a longliner to catch lancetfish which have eaten smaller ones that were already hooked. Because of their 'watery' muscles, it has been theorized that lancetfish are ambush predators; however, other aspects of their morphology such as the powerful tail may indicate that they are active predators.

Most mesopelagic fishes are generally not available to sport fishing; however, the pictured fish proves to be the exception to the rule. In fact, it is probably the only lancetfish so far caught by rod and reel, made even more remarkable by the fact that it was caught in broad daylight.

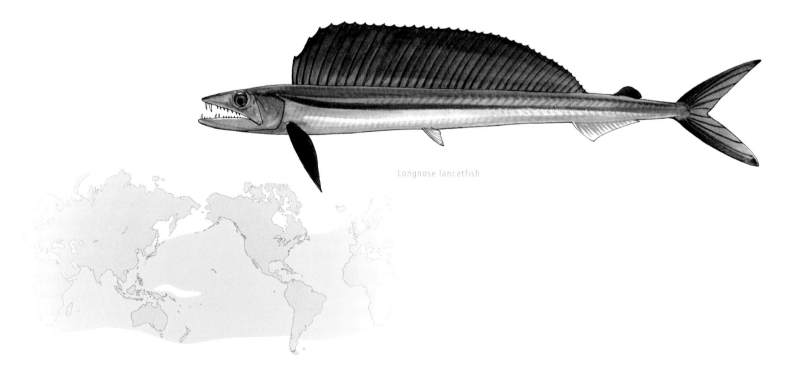

Longnose lancetfish

Rarely seen alive, this juvenile longnose lancetfish was caught on a research cruise off Hawaii and kept alive for studies of its eyesight.
Kerstin Fritsches

An extremely rare catch. This adult longnose lancetfish was caught in broad daylight under a fish aggregating buoy in Hawaii. The bait was set very deep. The fish was released in good condition.
David Itano

RAY'S BREAM
Brama brama

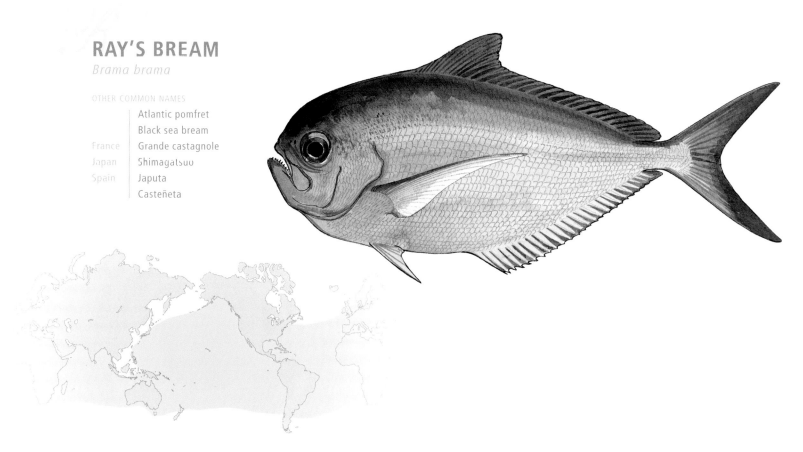

R ay's bream is officially named the Atlantic pomfret, but that name has not really been adopted throughout its range, hence the use of the more commonly used 'Ray's bream' here. This is a truly cosmopolitan pomfret, occurring widely in all three major oceans and in the Mediterranean. It is found throughout the water column of the open ocean, from the surface to depths over 1000 metres. This is due to its daily migratory habit of staying deep during the day and ascending at night with the so-called scattering layer rich in zooplankton, squid and the many other fishes which exist in this ecosystem. There are seven species of pomfret in the genus *Brama*, but this is by far the most widespread and commercially important. Ray's bream can be identified by its steeply arched forehead and an anal fin ray count of 30 or more.

Rather oddly, the UN Convention on the Law of the Sea classifies all members of the pomfret family, Bramidae, as highly migratory species, and while Ray's bream is often described as such, this appears to be pure speculation since there have been no specific studies of the movements of this important species.

Ray's bream is a relatively small pomfret, only growing to about 6 kg (13 lb), but it has a wide gape and fish as small as 2 kg (4.5 lb) freely take baited longline hooks designed to catch much larger fish such as tunas and billfish. It is caught by both pelagic longlining and bottom trawling, demonstrating its broad habitat preferences.

A relatively recent study of growth rates of Ray's bream in the Atlantic suggested that this species grows very slowly, with some fish of 2 kg (4.5 lb) being estimated at ten to 12 years old. The authors of the study admitted that these ages were not able to be validated, and noted that studies on the closely related Pacific pomfret (*Brama japonica*) indicated very fast growth rates and a short life span of perhaps three years.

Like other pomfrets, Ray's bream is an excellent table fish and a welcome bycatch of commercial fisheries. Historically, it was rarely taken by recreational anglers. However, in recent years, targeting by anglers of deep-water species such as hapuka (*Polyprion* spp.) over seamounts has seen catches of Ray's bream and other pomfrets increase.

SICKLE POMFRET
Taractichthys steindachneri

OTHER COMMON NAMES

	Monchong
Japan	Hirejiro-manzai-uo
Spain	Tristón segador

The sickle pomfret, probably better known as the monchong, especially in Hawaii, is a striking fish by any standard. It is covered with large oblong interlocking scales which have a markedly black, metallic sheen, reminding one immediately of a Japanese medieval warrior clad in black armor.

The sickle pomfret is widespread throughout the warmer waters of the Pacific and Indian oceans (its Atlantic counterpart is the almost identical long-fin pomfret, *Taractichthys longipinnis*). Other pomfrets, such as the similar looking lustrous pomfret (*Eumegistus illustris*) and Ray's bream (*Brama australis*), have much smaller dorsal and anal fins and smaller scales compared with the sickle pomfret.

This is one of the species which is part of fish communities over remote seamounts and is normally caught in depths of 300 metres or deeper. The sickle pomfret was largely unknown before long-lining techniques began to catch them in the 1960s. They are now regularly caught in association with bigeye tuna, albacore and lancetfish, almost invariably as a bycatch of longline campaigns which specifically target bigeye tuna by setting deep hooks. The pomfrets as a group are excellent table fish with pink to white, moist flesh, and the sickle pomfret is no exception. On Hawaiian markets, this species brings a premium price, being particularly popular in top-end seafood restaurants. As a result, sickle pomfret catches have been steadily increasing in Hawaii, prompting a small research program into the biology of this mysterious fish.

Sickle pomfret caught by longline average around 70 cm long (7 to 8 kg – 15 to 18 lb), but specimens up to 120 cm long and weighing 45 kg (99 lb) have been reported in some references. They apparently grow very quickly at first, reaching up to about 45 cm in their first year of life. However, unlike some other deep-sea fishes, they do not appear to be very long-lived, with the oldest specimen so far estimated at only nine years old. Scientists studying the species in Hawaiian waters have been intrigued to find that, even though this fish has tough armor-like scales, it is not impervious to attack from the ubiquitous cookiecutter shark.

FANFISHES

Pteraclis spp.,
Pteracombus spp.

OTHER COMMON NAMES
Wingfish
Spotted fanfish
Pacific fanfish
Japan | Benten-uo

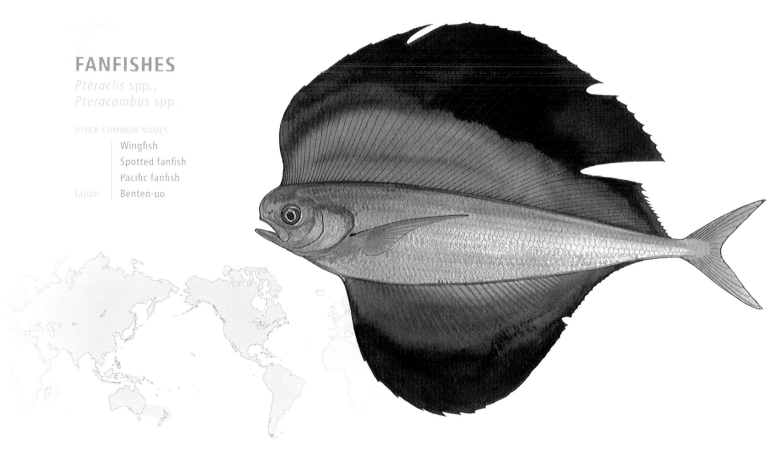

Spotted fanfish

Several species of the pomfret family Bramidae, the fanfishes or wingfishes, are among the more bizarre denizens of the open ocean. Both fanfishes and wingfishes show many general pomfret features such as a snub-nosed profile, upturned mouth and large scales, but their most obvious feature which sets them apart is their extraordinarily enlarged dorsal and anal fins. Amazingly, these fins are able to be folded completely into dorsal and ventral slots formed by rows of enlarged scales. As a result of the anal fin extending so far forward, the vent, or anus of the fanfishes is located under their throat.

The purpose of these huge fan-like fins can only be guessed at. It has been speculated that they may be raised suddenly to surprise potential predators, or that they may be used in sexual displays (like the tail of a peacock). Perhaps we will never know the true answer to this intriguing question.

The fanfishes are very rare, at least as far as their capture is concerned. Some specimens have been found washed up on beaches, but most have been taken by surface longline, indicating their pelagic habit. However, since these are small fish (maximum size about 60 cm) the chances of capture by this method, intended to catch much larger fish, must be quite slim.

There are three species of fanfish within the genus *Pteraclis*, all of which have the greatly expanded fins described above. They are the wingfish or spotted fanfish, *Pteraclis velifera*, a global species, the Pacific fanfish, *Pteraclis aesticola*, found only in the northwestern Pacific, and the Carolina fanfish, *Pteraclis carolinus*, confined to the Atlantic. Two other species of fanfish are known, both in the genus *Pterycombus*, and while the dorsal and anal fins of those species are large, they are not developed to anywhere near the size of the *Pteraclis* fanfishes.

Relatively few people have ever seen a living or just-caught fanfish, but descriptions by those who have suggest they are truly stunning. The huge fins of a live spotted fanfish are apparently washed with a vivid violet-blue and patterned with turquoise spots while the body is described as bright metallic silver. Photos of the Pacific fanfish show it to be equally spectacular, with intense deep blue-black fins and a metallic bronze-colored body.

OILFISH
Ruvettus pretiosus

OTHER COMMON NAMES

	Black oilfish
	Castor oil fish
France	Rouvet
Japan	Baramutsu
Spain	Escolar clavo

The oilfish is a member of the snake mackerel family, Gempylidae. This is the largest of the 23 known gempylid species, growing to a length of 2 metres, and a maximum recorded weight of 63.5 kg (140 lb). It is a highly distinctive fish identified by its brown to black color, its tuna-like shape and its sharp, bony tubercles between the scales, especially prominent along the mid-belly. The main species with which the oilfish is often confused is the escolar, *Lepidocybium flavobrunneum*. The two can be distinguished by the prominent wavy lateral line of the escolar, its lack of spiky processes between the scales and a triple keel on the caudal peduncle, which the oilfish lacks.

The oilfish has a broad, albeit patchy global distribution in tropical and temperate waters. It occurs on each side of the three major oceans, and around many islands and archipelagos in between. Regarded as a benthopelagic species, it is primarily caught as a bycatch of long-lining, usually at depths greater than 200 metres, ascending higher in the water column at night, and descending to the bottom during the day.

Polynesians have traditionally fished for oilfish (palu) using a specially designed wooden 'palu hook'. In fact, a publication by Goodger in 1928 imputed the geographic distribution of this species in the Pacific by recording the islands where this distinctive type of hook was used.

The oilfish has been given its name for very good reasons. Its flesh, which is beautifully white and dense when cooked, has a very high oil content and, like many other deep-water fishes, also contains waxy esters. The fish is perfectly edible – in fact, delicious – but the waxy ester is a highly effective purgative, or laxative. Even though the oilfish has this rather unpleasant reputation, it is still keenly sought by those who consider the flavor of the flesh worth the risk of the aftermath. The flesh of another gempylid fish, the escolar, described elsewhere, is not as oily as that of the oilfish, which may explain why both species are often marketed under the name 'escolar', perhaps to hide the fact that one might be buying oilfish.

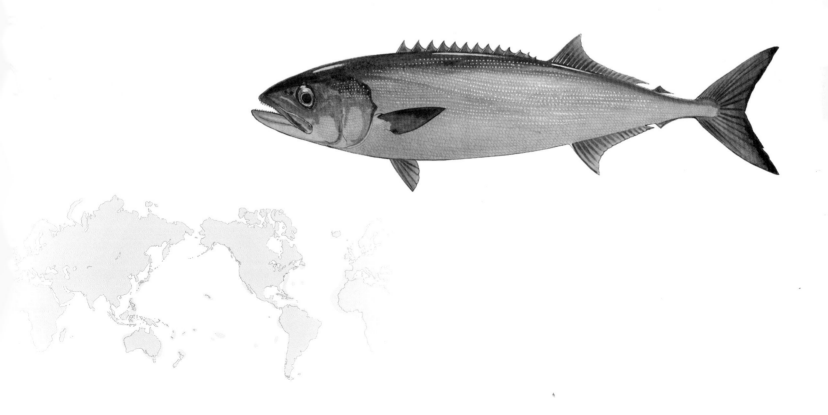

ESCOLAR
Lepidocybium flavobrunneum

OTHER COMMON NAMES

France	Escolier noir
Japan	Aburasokomutsu
Spain	Escolar negro

The escolar is a member of the snake mackerel family, Gempylidae. It is a robust fish, more tuna-like in shape than the more laterally compressed form of most other members of the family.

It is a global species, occurring in all three major oceans, although it is absent from the Mediterranean. Its broad distribution ranges from cool temperate latitudes right through to the tropics. The escolar is a mesopelagic (midwater) fish which migrates to near the surface at night. It is a common bycatch species of the oceanic longline fisheries, in some cases being an important source of additional income due to its market acceptability as a prime table fish.

The escolar is most commonly confused with the oilfish (*Ruvettus pretiosus*) since both are quite similar looking brown to black tuna-like fish and are often caught together. Several features, however, immediately separate the two. The oilfish is very rough to the touch, with sharp projections (tubercles) between the scales, which the escolar lacks. This is especially the case along the midline of the belly. The escolar has a very obvious wavy lateral line, while that of the oilfish is obscure and the escolar possesses a prominent caudal keel, flanked by two smaller keels, whereas the oilfish has no keel.

Some minor differences in morphology between Atlantic and Indo-Pacific escolar led to a recent genetic study of fish from the two regions. This found distinct differences in the DNA of fish from each region, suggesting the recognition of two species, or at least subspecies, of escolar.

As is the case with many of the fishes which dwell in midwater, the biology of the escolar is poorly understood. Based on changes in seasonal catches, it is thought to migrate from productive feeding areas in cooler latitudes to more tropical areas where it spawns. Larvae have been mainly found near oceanic islands, suggesting that these are the areas where spawning takes place.

The maximum size to which escolar grow is perhaps 2 metres in length and 43 kg (95 lb); however, fish of that size would undoubtedly be very rare.

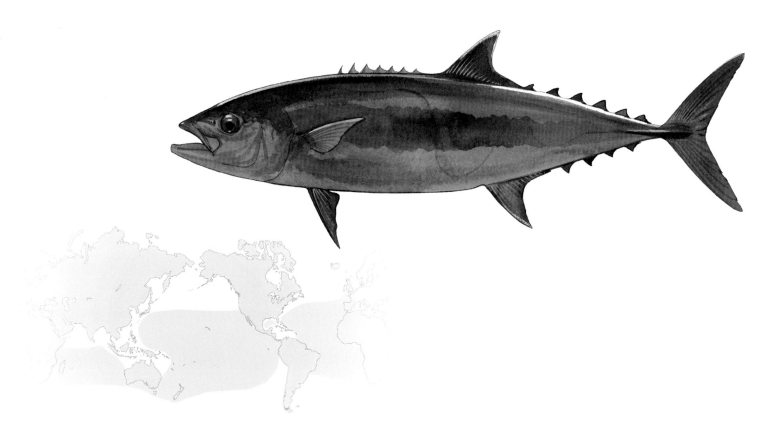

SNAKE MACKEREL
Gempylus serpens

OTHER COMMON NAMES

France	Escolier serpent
Japan	Kurotachikamasu
Spain	Escolar de canal

The snake mackerel gives its name to the family, Gempylidae. It is an elongated, sinuous fish with well set back dorsal and anal fins and a series of finlets in front of a broad, forked tail. At the other end, it has a long snout and a large mouth equipped with very large, sharp teeth. Fishes with these features, which include barracuda and needlefish, are 'hover predators' – meaning that when they detect prey, they remain motionless before making a sudden forward lunge to grab the prey in their long, toothed jaws.

This is probably the most widespread of the gempylids. It occurs globally in tropical, subtropical and, less frequently, temperate waters. In contrast to many of the gempylids, which are benthopelagic, the snake mackerel is often found in the surface layers of the world's oceans and is a common bycatch of oceanic longline fisheries. Catches of intermittent, single fish on a longline suggest that this is a solitary hunter.

The snake mackerel grows to a length of at least 1.2 metres.

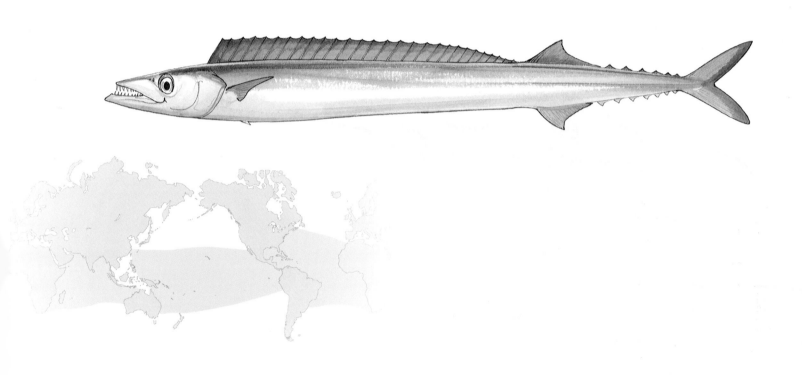

OPAH

Lampris guttatus,
L. immaculatus

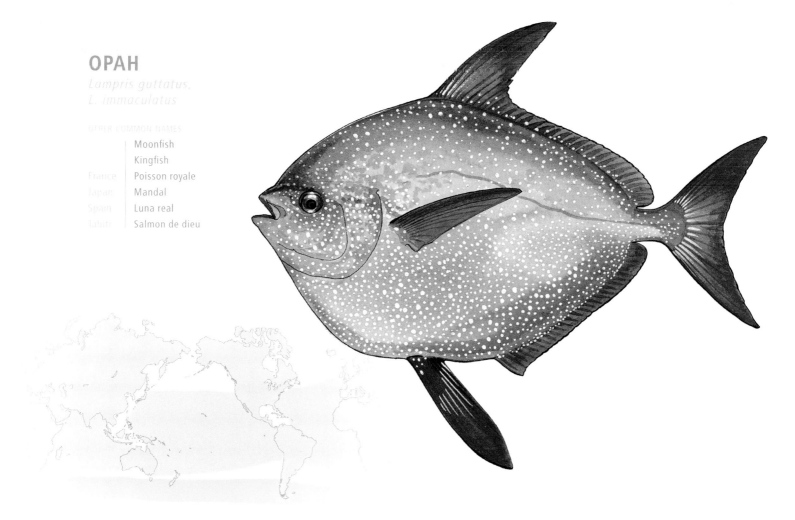

☐ Opah
◼ Southern opah

The opah is one of the most striking inhabitants of the open ocean. It belongs to a group of fishes (Lampriformes) which are amazing in their bizarre shapes and adaptations. The oarfish is probably the best known member of this group, its most obvious similarity to the opah being the brilliant orange-rose coloration of the fins. The reason for this general color pattern in fishes of such differing body forms which live in mid-ocean depths, but with presumably very different lifestyles, is not known, although it is likely due to orange and red colors appearing as black at depth.

The opah swims quite rapidly by flapping its pectoral fins like the wings of a bird (reminiscent of a swimming penguin). To achieve this, the muscle supplying these fins is extremely well developed, and it is this muscle which is considered a great delicacy. The Polynesians hold opah in great reverence, their name for the fish literally meaning 'food of the gods'.

There are two species of opah, the most common of which, *Lampris guttatus*, is illustrated. Opahs are characterized by a very deep, laterally compressed body and greatly elongated pelvic fins. The common opah is particularly colorful, with silvery-pink sides covered with silver spots, and crimson or vermilion fins. It is worth recounting an early glowing description of the opah by Robert Harrison in 1769:

> ... all fins and the tail are of a fine scarlet; but the colours and beauty of the rest of the body, which is smooth and covered with almost imperceptible scales, beggars all description: the upper part being a bright green, variegated with whitish spots, and enriched with a golden hue, like the splendour of a peacock's feather: this by degrees vanishes in a bright silvery, and near the belly the gold again predominates in a lighter ground than on the back.

In contrast, the southern opah is quite plain, although it does possess the characteristic crimson fins and tail.

The common opah is distributed through the warmer waters of all oceans, including the Mediterranean. It occupies depths from the surface to at least 350 metres. The southern opah lives in much colder water and has a circumglobal distribution in the southern hemisphere in a band between latitudes 40°S and 60°S.

Very little is known about movements of opah. Several common opah have been tagged with popup satellite tags off Hawaii, some of which moved considerable distances. It was also shown that opah dive as deep as 400 metres during the day, but like many other open-ocean fishes, ascend to near surface at night.

Juvenile opah are much more streamlined than adults. As well, they lack the enlarged pelvic fins of the adult. Growth rates are not well understood, but are likely to be rapid, since large adults would appear to be relatively common in relation to other sizes. A recent Hawaiian study estimates an age range of one to six years for fish between 20 kg (44 lb) and 70 kg (155 lb). Stomach content studies have revealed an extremely wide variety of food items, including fish, squid, crustaceans and jellyfish (and plastic bags). The maximum size of the common opah quoted in the literature is 275 kg (605 lb). However, this is very unlikely since no opah over 90 kg (200 lb) has been recorded in the Hawaiian longline fishery, despite over 30 000 having been measured. The southern opah is a much smaller species, growing only to around 30 kg (66 lb).

Like other large oceanic fishes, opah produce huge quantities of eggs, presumably due to the precarious environment into which they are shed. However, spawning behavior, seasons and areas are unknown.

Opah are caught incidentally in deep-set tuna longline fisheries, and bring a good price on some markets. Depths of hooking are normally between 100 and 250 metres, but opah may also be caught by drift nets set at the surface at night.

There are scattered records of the species caught by recreational anglers. A 58 kg (128 lb) specimen was caught in 1973 by an angler off Penzance, England, while the all-tackle record appears to be an opah weighing 74 kg (163 lb) caught off Port San Luis, California, in 1998. In an unusual occurrence, in June 2008, three opah were caught on separate occasions by recreational anglers off New South Wales, Australia, all weighing about 55 kg (120 lb).

OARFISH
Regalecus glesne

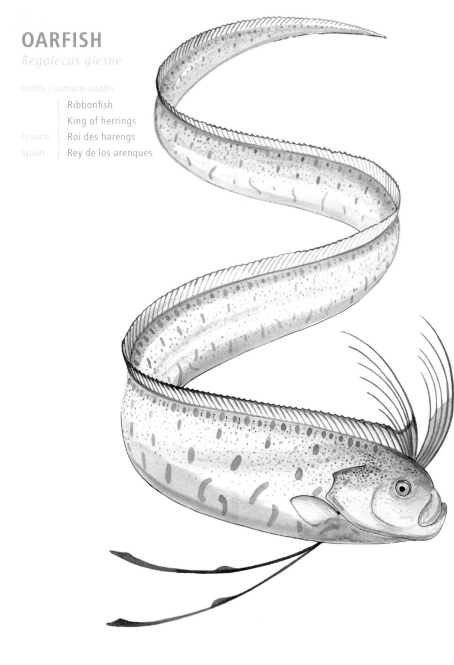

If there is one creature that might well explain some of the reports of mythical sea serpents in the past, it would have to be the oarfish. This is a most unusual, mainly mesopelagic (midwater) fish which, on its rare sightings at the surface or washed up on beaches, must have been the stuff of nightmares to ancient mariners.

The oarfish derives its name from the two extremely long, thin pelvic fins. The ends of these fins expand into a roundish membrane, giving the impression of two long oars. However, while the function of these stick-like fins is unknown, they are almost certainly not used in any powered locomotory stroke. Even though oarfish are rarely seen alive, footage of one filmed in the Bahamas revealed its extraordinary method of swimming. This showed that the normal position of the oarfish when feeding is hanging vertically, head facing the surface, with its long dorsal fin rippling almost hypnotically in a continuous traveling wave. The other rather odd common name of the oarfish, the 'king of herrings', is a translation of its Latin name, *Regalecus*. This was bestowed over 200 years ago, and was probably due to the oarfish's possession of a crest of very long, crimson rays at the beginning of the first dorsal fin, giving the appearance of a flamboyant royal crown.

The oarfish has a global distribution, often revealed by the location of stranded specimens, but also by occasional captures on longline or even by recreational anglers. Oarfish are thought to feed mainly on small shrimps, salps and squid using their highly protrusible mouth to suck in food items in great gulps.

The oarfish, together with two other obscure species, belongs to the family Regalecidae which, in turn, is one of seven families within a larger group of fishes called the Lampriformes. These poorly understood fishes include such unusual species as the opahs, crestfishes, ribbonfish, dealfish, tapertails and band-fish. Many of these are mesopelagic fishes, meaning that they neither live at the surface nor on the bottom, but inhabit the midwater twilight zone of the open ocean. Apart from their habi-tat, the other feature which most of these fishes have in common is their coloration – usually possessing brilliant orange to crimson fins mounted on silvery bodies. Usually these fishes are found washed up on beaches, so their living colors are rarely seen. However, photographs of freshly stranded fish show a stunning array of unexpected reflective metallic colors along the flanks of the oarfish, ranging from silver to violet.

The oarfish is often cited as the longest bony fish that swims, but the true maximum size to which it grows is shrouded in mystery. Lengths of 11 to 15 metres are often referred to (55 feet is commonly cited), but with no proof of those sizes, they must remain dubious. Two photos which I have examined show very large oarfish being held by rows of men, the first shown here and the second showing an even larger fish held by 15 US marines. Estimating the lengths of these fish by comparison with objects in the photos suggests these fish would 'only' be about 5 metres and perhaps 6.5 metres long. It would therefore seem that a fish measuring 11 metres, let alone one of 15 metres, would be somewhat unlikely.

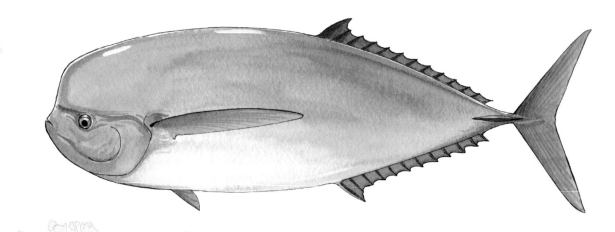

LOUVAR
Luvaris imperialis

The louvar, *Luvarus imperialis* – also called luvar or luvalu – is a single world-wide species and the only member of its own family, Luvaridae. Growing to at least 150 kg (330 lb) and in life, colored bright pink with orange fins, this impressive fish is only rarely seen, let alone caught by anglers or commercial fishers. Apart from the fact that it happens to be a naturally rare, solitary fish, perhaps the main reason that the louvar is rarely taken on baited hooks is that its diet consists almost entirely of jellyfish and other gelatinous macroplankton called cnidophores.

Even though most known specimens of louvar have been found washed up or stranded on beaches, this is a true open-ocean species that spends most of its time well away from land, probably in mid-water. Its bright pink-orange coloration is similar to other strange fishes of this twilight world such as the oarfish and the opah, which strongly suggests that it dwells in relatively deep water where orange and pink appear black or gray, helping such fish to merge with the shadowy background.

The louvar was long thought to be related to the tunas, primarily on the basis of its tuna-like tail and narrow tail wrist (caudal peduncle) with promi-

nent lateral keels. However, more recent studies of juvenile and larval specimens clearly indicate that the most likely close relatives of the louvar are in fact the surgeon fishes (family Acanthuridae) – much smaller reef-dwelling fish equipped not with keels, but with razor sharp 'blades' on their tail wrist. This is particularly interesting since another large oceanic fish, the huge sunfish, has been found to be very closely related to the pufferfishes. Two cases of small, inshore fishes with very large oceangoing relatives.

Global records of louvar are quite scattered, but piecing together information from many sources shows that they are probably most common in the Mediterranean, followed by the Pacific coasts of the United States and Mexico. Occasional specimens have been netted or washed up in England, while adult fish have also been washed up or caught along the Australian southeast coast. I am aware of three recorded captures of louvar on rod and reel. The first was a 50 kg (110 lb) specimen caught in 1995 off Baja California, while the second and third, weighing 80 kg (176 lb) and 112 kg (246 lb), were caught off New South Wales, Australia, in 1997 and 2009.

RAINBOW CHUB

Sectator ocyurus

The sea chubs, also known as drummers or buffalo bream, are members of the family Kyphosidae. Most species within the family are coastal; however, some associate strongly with floating objects and, as a result, become part of slow-drifting open-ocean ecosystems.

Some species are more oceanic than others, the rainbow chub being one species which is a permanent resident of that environment. This striking looking fish is found mainly in the eastern Pacific, from Baja California to Peru, but also around Hawaii and the Society Islands and islands off southern Japan. They are frequently seen in numbers under floating logs or debris far from land.

This is an unmistakable fish. It is slightly more elongated than most chubs, and marked with bright blue and yellow stripes longitudinally along the body. Its teeth form single rows in either jaw, each tooth said to resemble the shape of a golf club (presumably the blade of an 'iron').

The rainbow chub grows to about 60 cm and 2 kg (4.5 lb) in weight.

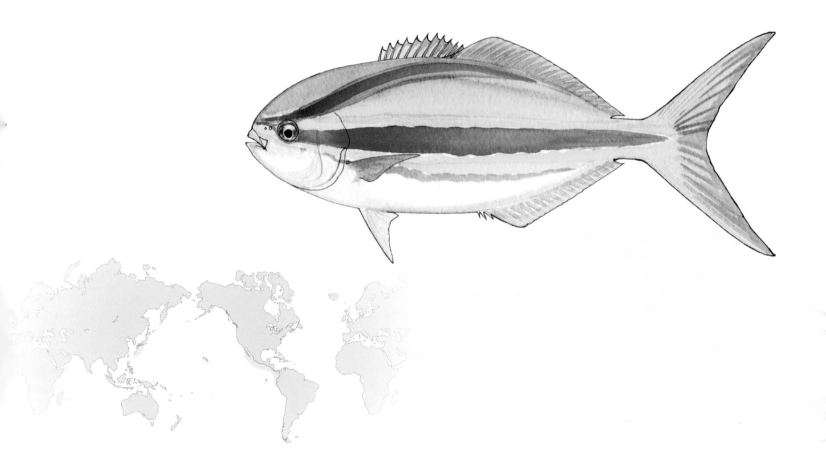

CORTEZ SEA CHUB
Kyphosus elegans

	Cortez chub
France	Calicagère de Cortez
Spain	Chopa de Cortez

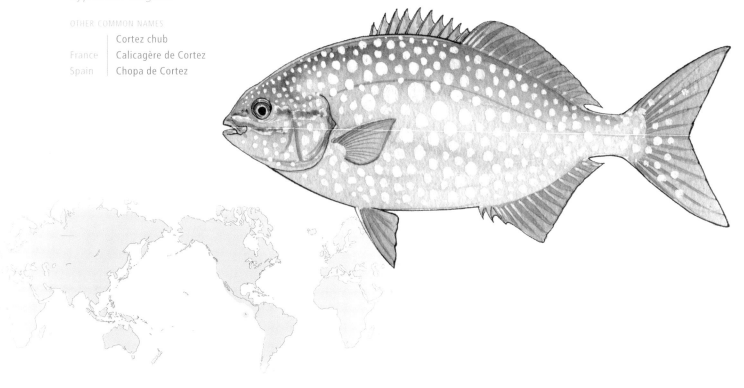

The Cortez sea chub is the other species of the family Kyphosidae selected for inclusion, partly because of its habit of associating with floating objects, and also because of its 'classic' kyphosid appearance. Like the rainbow chub, the Cortez sea chub is, at times, a component of drifting ecosystems of the open ocean. This species occurs only in the eastern Pacific, from southern California right along the central American coast. It also occurs, not surprisingly, in the Sea of Cortez (Gulf of California) and many offshore islands in the region, including the Galapagos.

The Cortez sea chub is colored uniformly blue-gray, often with fairly regular paler blue spots covering the whole body. The spots may be transient and fade after capture. It also shows the many horizontal rows of scales characteristic of members of the family. The Cortez sea chub grows to around 50 cm and a weight of 2 kg (4.5 lb).

TRIPLETAIL
Lobotes surinamensis,
L. pacificus

Until recently, the tripletail was considered a single worldwide species, *Lobotes surinamensis*. However, the population in the eastern Pacific is now considered to be a separate species, *Lobotes pacificus*, albeit, with only subtle differences between the two. These two species are the only members of the family Lobotidae, which are very similar in appearance to the freshwater tripletails (family Datnioididae). This similarity even extends to sharing the behavior of laying on their side and floating like leaves. It is therefore likely that one evolved from the other, most probably the saltwater forms evolving from freshwater ancestors.

This is a distinctive fish which bears little resemblance to any other species found in the same oceanic habitat. They are laterally compressed robust fish with mottled light and dark brown coloration. Their second dorsal, anal and caudal fins resemble a single three-lobed tail.

Juvenile tripletails are differently marked from adults, being pale yellow with black streaks. They also drift on their sides, and as such, closely resemble floating dead leaves of mangroves, with which they often associate.

The common tripletail occurs in all three major oceans, but as noted, is replaced in the eastern Pacific by the Pacific tripletail. Locations of recorded captures tend to be coastal, but tripletail are also commonly recorded near isolated islands (for example, Fiji and Tuvalu) and are commonly seen under weedlines and floating debris far from land.

The method of movement and rate of mixing of tripletails across large distances must be efficient since they are so widespread. Rather than by means of actively swimming, however, it is most likely that dispersal is achieved passively, by fish simply staying under floating logs or other debris carried on ocean currents.

The largest tripletail on record weighed 19.2 kg (42.3 lb) and was caught off Zululand, South Africa, in 1989. The next largest appears to be one caught off Florida in 1998 which tipped the scales at 18.5 kg (40.7 lb).

■ Tripletail
■ Pacific tripletail

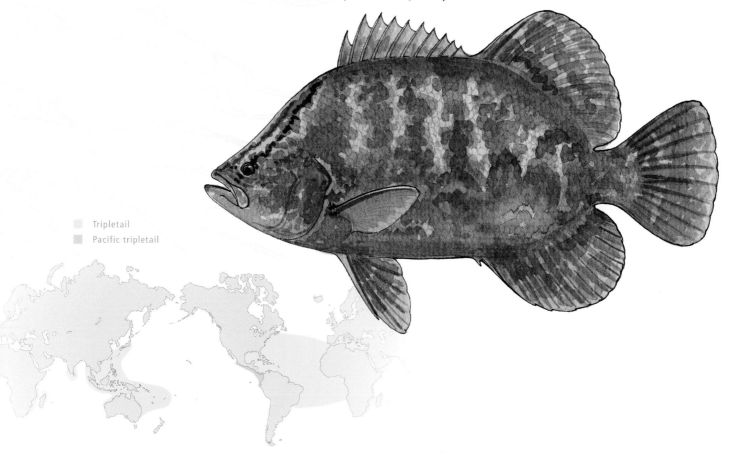

UNICORN LEATHERJACKET
Aluterus monoceros

OTHER COMMON NAMES

	Pelagic leatherjacket
	Unicorn filefish
France	Bourse loulou
Japan	Usubahagi
Spain	Lija barbuda

The leatherjackets and filefishes, members of the large family Monocanthidae, are nearly all inshore or estuarine fishes. However, some, including the unicorn leatherjacket, have entered the pelagic environment via their habit of associating with floating objects such as logs and kelp.

Presumably as a result of this lifestyle, this is one of very few species of leatherjacket that has a global distribution. It is found on both sides of all three major oceans, and around a number of oceanic islands. Juveniles often associate with jellyfish, while adults commonly associate with floating weeds, algae and debris.

The unicorn leatherjacket can be identified by its upturned mouth and its dull gray to brown color covered on the dorsal surface by numerous small brown spots.

This is a relatively large monocanthid, growing to about 75 cm and a weight of at least 2 kg (4.5 lb).

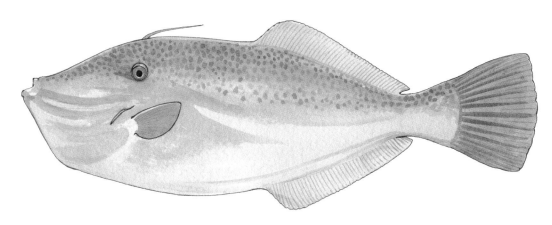

OCEANIC PUFFER
Lagocephalus lagocephalus

OTHER COMMON NAMES

	Pelagic toadfish
	Oceanic toadfish
France	Compère lièvre
Japan	Kumasakafugu
Spain	Tamboril liebre

As is the case for many of the species of fish which have successfully colonized the specialized habitat of the surface layers of the open ocean, the oceanic puffer is very widespread, having a global distribution. It occurs in all three major oceans as well as the Mediterranean.

The coloration of the oceanic puffer reflects its pelagic existence. It displays the classic dark blue-black dorsal surface and pearly white sides and belly, similar to the pattern of so many fishes of this zone, regardless of their phylogenetic origins. The oceanic puffer is also more elongated and has a broader tail than its inshore relatives, suggesting an active swimming lifestyle. This is quite a large pufferfish, growing to more than 60 cm in length and over 3 kg (7 lb) in weight.

As is the case with many members of the pufferfish family, Tetraodontidae, the flesh and organs of the oceanic puffer are poisonous. The celebrated 18th-century naval captain James Cook was poisoned after eating the liver of a fish, almost certainly an oceanic puffer, in New Caledonia. He survived, but only after three days of sickness and delirium.

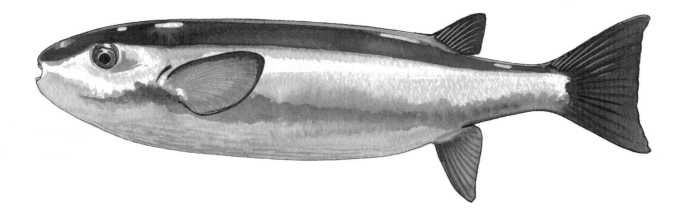

OCEAN SUNFISH
Mola mola

The word 'mola' is Latin for 'millstone', an appropriate name to bestow on this huge, almost disc-shaped fish. The ocean sunfish does not have a tail at all, but rather a fusion of dorsal and anal fin rays at the back of the rounded body to form a leathery structure called a 'clavus'. The name sunfish not only refers to their round shape and reflective surface, but also to the fact that these fish often lie on their sides at the surface, apparently sunning themselves. The sunfishes are quite closely related to the pufferfishes (also known as blowfish or toadfish), sharing features such as fused beak-like teeth plates and a small mouth. However, unlike their relatives, sunfish are not poisonous to eat and, in fact, are considered a delicacy in some countries.

One of the oft-quoted facts about sunfish is that they mainly eat jellyfish and salps, and while this is certainly true, it is now clear that their diet is much more varied. Occasionally, anglers report sunfish taking live fish baits being trolled or drifted for tuna or billfish. In captivity, sunfish will readily eat a wide range of offered food, including crustaceans and fish.

There are four species of sunfish belonging to the family Molidae. These are the ocean sunfish, *Mola mola*, the Southern Ocean sunfish, *Mola ramsayi*, the slender sunfish, *Ranzania laevis*, and the sharptail mola, *Masturus lanceolatus*. The sharptail sunfish can be identified by a rearward extension in the middle of the

An oceanic sunfish
being cleaned
over a deep reef
slope by longfin
bannerfish. The
sunfish is the
largest bony fish.
*marinethemes.com/
Jez Tryner*

clavus. The Southern Ocean sunfish lacks a distinct band of smaller denticles at the base of the clavus, which is present in the oceanic sunfish. The slender sunfish is longer than it is deep, has very slender dorsal and anal fins and its body is patterned with wavy, dark-edged stripes and vertical broken bands. The slender sunfish is the only one of the group which possesses a vertical mouth slit.

All four species of sunfish are primarily oceanic, occurring in tropical and subtropical waters of all three major world oceans. While adults tend to occur offshore, juvenile fish often come close to the coast where their large dorsal fin often causes them to be mistaken for sharks. Sunfishes tend to be solitary or to swim in small groups, but one species, the slender sunfish, is known to school and is sometimes involved in mass strandings of up to 200 individuals.

The ocean sunfish is often cited as the world's largest bony fish, with a weight of more than 2000 kg (4400 lb) often quoted. It could be, though, that this figure is based on a single fish – a famous specimen which was 'landed' in bizarre circumstances early last century. On 18 September 1908, the steamship *Fiona* hit an object 40 miles off Sydney Heads with an almighty shudder, causing the vessel to come to a halt. To the amazement of the captain and crew, an inspection revealed a massive fish (they first thought it was an elephant!) trapped in the port propeller. The ship limped back to port on its remaining screw, whereupon the great fish was hoisted clear of the water, revealing a huge sunfish. Good photos of the fish were taken, as well as measurements – 10 feet 2 inches long (310 cm), 13 feet 4 inches (406 cm) from the tips of the dorsal and anal fins – but unfortunately, the fish was not weighed. Rather, the Captain and the winchman of the *Fiona* estimated its weight to be 'about 2 tons', a figure which now seems to have entered the popular and scientific literature as a true weight. In any case, the photo of that massive fish lends credibility to the general acceptance of the sunfish being the largest bony fish on the planet.

Sunfish growth rates are probably very rapid. A juvenile *Mola mola* which was kept in the famous million-gallon open-ocean tank at the Monterey Bay Aquarium in California grew from 27 kg (60 lb) to an amazing 408 kg (900 lb) in just 15 months. In fact, it grew so big that it was eventually lifted by helicopter to be released in the ocean.

BIBLIOGRAPHY

While many scientific papers and reports were invaluable in the preparation of this book, it would be superfluous to list all of those here. Rather, the interested reader is directed to the following more accessible general works which were also of considerable assistance in compiling this work.

Allen, G 1997, *Marine Fishes of Tropical Australia and South-east Asia*, Western Australian Museum, Perth.

Allen, LG, DJ Pondella & MH Horn 2006, *The Ecology of Marine Fishes: California and adjacent waters*, University of California Press, Berkeley.

Anon. 2003, Proceedings of the Third International Billfish Symposium, Cairns, Queensland, Australia, 19–23 August 2001, published as *Marine & Freshwater Research* 54 (4), 287–584.

Anon. 2009, *World Record Game Fishes*, International Game Fish Association, Florida.

Bullen, FT 1904, *Creatures of the Sea*, The Religious Tract Society, London.

Carrier, JC, JA Musick & MR Heithaus (eds) 2004, *Biology of Sharks and their Relatives*, CRC Press, Boca Raton.

Castro, JI 1983, *The Sharks of North American Waters*, Texas A & M University Press, College Station.

Collette, BB & CE Nauen 1983, FAO species catalogue, vol. 2, Scombrids of the world: An annotated and illustrated catalogue of tunas, mackerels, bonitos and related species known to date, *FAO Fisheries Synopsis*, 125 (2), FAO, Rome.

Collette, BB, JR McDowell & JE Graves 2006, Phylogeny of recent billfishes (Xiphioidei), *Bull. Mar. Sci.* 79 (3): 455–68.

Compagno, LJV 1984, Sharks of the world: An annotated and illustrated catalogue of shark species known to date, vol. 4, part 2, Carcharhiniformes, *FAO Fisheries Synopsis*, no. 125, vol. 4, part 2, FAO, Rome.

Compagno, LJV 2001, Sharks of the world: An annotated and illustrated catalogue of shark species known to date, volume 2, Bullhead, mackerel and carpet sharks (Heterodontiformes, Lamniformes and Orectolobiformes), *FAO Species Catalogue for Fishery Purposes*, no. 1, vol. 2, FAO, Rome.

Compagno, L, M Dando & S Fowler 2005, *A Field Guide to the Sharks of the World*, Collins, London.

Dunn, B & P Goadby 2000, *Saltwater Game Fishes of the World: An illustrated history*, Australian Fishing Network.

Eschmeyer, WN, ES Herald & H Hammann 1983, *A Field Guide to Pacific Coast Fishes of North America*, Houghton Mifflin Company, Boston.

FAO (Food and Agriculture Organization of the United Nations) 1998–2001, The Living Marine Resources of the Western Central Pacific, *FAO Species Identification Guide for Fishery Purposes*, vols 2–6, FAO, Rome.

FAO (Food and Agriculture Organization of the United Nations) 2002, The Living Marine Resources of the Western Central Atlantic, *FAO Species Identification Guide for Fishery Purposes*, vols 1–3, FAO, Rome.

Gray, Z 1926, *Tales of the Angler's Eldorado: New Zealand*, Hodder & Stoughton, London.

Helfman, GS, BB Collette & DE Facey 1997, *The Diversity of Fishes*, Blackwell Science, Malden, MA.

Holder, CF 1903, *The Big Game Fishes of the United States*, The Macmillan Company, New York.

Holland, KN, ML Domeier, DB Holts, JE Graves & JG Pepperell (eds) 2006, Proceedings of the Fourth

International Billfish Symposium, Avalon, Santa Catalina Island, California, 31 October – 3 November 2005, published as *Bulletin of Marine Science*, 79 (3), 429–874.

Horsman, PV 1985, *The Seafarer's Guide to Marine Life*, Croom Helm, London and Sydney.

Klimley, AP & DG Ainley (eds) 1996, *Great White Sharks: The biology of Carcharodon carcharias*, Academic Press, San Diego.

Kohler, NE, JG Casey & PA Turner 1995, Length–weight relationships for 13 species of sharks from the western North Atlantic, *Fish. Bull.*, 93: 412–18.

Laboute, P & R Grandperrin 2000, *Poissons de Nouvelle-Caledonie*, Editions Catherine Ledru, Nouméa.

Last, PR & JD Stevens 1994, 2009, *Sharks and Rays of Australia*, CSIRO, Australia.

McClean, AJ & K Gardner 1984, *McClane's Game Fish of North America*, Times Books, New York.

Mather, CO 1976, *Billfish: Marlin, broadbill, sailfish*, Saltaire Publications Ltd, Sidney BC, Canada.

Nakamura, I 1985, FAO species catalogue, vol. 5, Billfishes of the world: An annotated and illustrated catalogue of marlins, sailfishes, spearfishes and swordfishes known to date, *FAO Fisheries Synopsis*, 125 (5), FAO, Rome.

Norman, JR & FC Fraser 1937, *Giant Fishes, Whales and Dolphins*, Putman, London.

Parin, NV 1968, *Ichthyofauna of the Epipelagic Zone*, Acad. Sci. USSR., Inst. Oceanol.; trans. US Dept Commerce, Fed. Sci. Tech. Insf., Springfield VA.

Paulin, C, C Roberts, A Stewart & P McMillan 2001, *New Zealand Fish: A complete guide*, Te Papa Press, Wellington.

Paxton, JR & WN Eschmeyer (eds) 1994, *Encyclopedia of Fishes*, UNSW Press, Sydney.

Pepperell, JG (ed.) 1992, *Sharks: Biology and fisheries*, CSIRO, Australia.

Randall, JE, GR Allen & RC Steene 1990, *Fishes of the Great Barrier Reef and Coral Sea*, Crawford House Press, Bathurst.

Roughley, TC 1951, *Fish and Fisheries of Australia*, Angus & Robertson, Sydney.

Smith, MM & PC Heemstra 1986, *Smiths' Sea Fishes*, Springer-Verlag, Berlin.

Stead, DG 1933, *Giants and Pigmies of the Deep*, The Shakespeare Head Press, Sydney.

Stevens, JD (consulting editor) 1987, *Sharks: An illustrated encyclopedic survey by international experts*, Golden Press, Sydney.

Tinsley, JB 1964, *The Sailfish: Swashbuckler of the open seas*, University of Florida Press, Gainsville.

Tricas, TC, K Deacon, P Last, JE McCosker, H Walker & L Taylor 1997, *Sharks & Rays*, The Nature Company Guides, Time Life Books.

Van der Elst, R 1993, *A Guide to the Common Sea Fishes of South Africa*, Struik, Cape Town.

Useful websites

www.austmus.gov.au/fishes/ (The Australian Museum, Ichthyology Department)

www.elasmo-research.org (ReefQuest Center for Shark Research)

www.fao.org (Food and Agriculture Organization of the United Nations)

www.fishbase.org (Fishbase website)

www.flmnh.ufl.edu/fish/ (Florida Museum of Natural History)

www.iattc.org (Inter-American Tropical Tuna Commission)

www.iccat.int (International Commission for the Conservation of Atlantic Tunas)

www.nmfs.noaa.gov (US National Marine Fisheries Service)

www.spc.int/oceanfish/ (Secretariat of the Pacific Community, Oceanic Fisheries Programme)

www.topp.org (Tagging of Pelagic Predators project)

www.wcpfc.int/ (Western and Central Pacific Fisheries Commission)

INDEX

man-of-war fish 237
manta ray 5, 216, **220**, 220–221, 225, 226
marlin, black 5, 22, 28, 29, 39, 41, **43**, 43–47, 61, 99
    blue 5, 6, 14, 27, 28, 34, **39**, 39–42
    striped 20, 24, 31, 35, **48**, 48–52
    white 42, **53**, 53–54, 61, 62
Megalops atlanticus 161
    cyprinoides 161
Mediterranean spearfish 60, 62, 63, 64
Mobula 220
Mola mola 39, 256, 258
Molidae 256
monchong 241
moonfish 246

narrow-barred Spanish mackerel 97, **114**, 114–116, 119
Naucrates ductor 148
needlefish, flat **232**, 233
    hound **232**
    keeled **232**
needlefishes 232–233
Nematistiidae 166
Nematistius pectoralis **166**
Nomeidae 237
Nomeus gronovii 237
northern bluefin tuna 81
northern Pacific saury 231

oarfish 246, **248**, 248–249
ocean sunfish 3, **256**
oceanic puffer **255**
oceanic whitetip shark 3, 4, 195, **196**, 197, 202
Odontapsidae 187
oilfish 243, 244
    black **243**
Oncorhynchus tshawytscha 169
ono 111
opah 4, 32, 212, **246**, 246–247, 249, 250
    southern 246, 247
Orcynopsis unicolour 100, 102
oxeye herring 161

oriental bonito 90, 100, 101

Pacific bluefin tuna 34, 81, 85
Pacific sardine **235**
Pacific tripletail 253
pelagic stingray **219**
pelagic thresher shark **186**, 187
pelagic toadfish **255**
permit 5, **143**, 143–144
Phtheirichthys lineatus 225
pilotfish **148**, 197
plain bonito 102
Platybelone argalus platyura 233
Pomatomidae 165
pomfret, longfin 241
    lustrous 241
Pomatomus saltatrix **165**
pompano 143, 144, 145
pompano dolphinfish 150, 152
porbeagle shark 168, 174, 178, **182**, 182–183, 184, 187
Prionace glauca 208
Pseudocaranx dentex 136
Pseudocarcharias kamoharai 211
Pteraclis aesticola 242
    velifera 242
Pterothrissu 163
Pteroplatytrygon violacea 219
puffer, oceanic **255**

queenfish, deep 145
    double-spotted 145
    giant **145**
    needle-scaled slender 145

Rachycentridae 159
Rachycentron canadum 225, 226
rainbow runner **146**, 147
Ranzania laevis 256
Rastrelliger brachysoma 107
    faughni 107
    kanagurta 107
ray, devil 220–221
    eagle **222**, 222–223
    manta 5, 216, **220**, 220–221, 225, 226

pelagic **219**
Ray's bream 8, 32, 66, **240**
red jack **140**
Regalecidae 248
Regalecus glesne 248
remora, common **225**, 225–226
    spearfish 225, 226
Remora australis 225
    brachyptera 225
    remora **225**
Remorina albescens 225
requiem shark 195
Rhincodon typus 204, 214
ribbon halfbeak 234
roosterfish **166**
roundjaw bonefish 163
roundscale spearfish **60**, 62, 63
Ruvettus pretiosus 243

sailfish 3, 10, 29, 31, 35, 55–59, 61
Salmo salar 167
salmon, Atlantic 4, **167**, 167–168, 170
    chinook 4, 167, **169**, 169–171
salmon shark 21, 170, 173, 174, 177, 178, 182, 183, **184**, 184–185, 211
Salmonidae 167
samson fish 123, **128**, 128–129
sandbar shark 195, 199, **200**, 201
Sarda australis 100, 101
    chiliensis chiliensis 101
    chiliensis lineolata 101
    orientalis 101
    sarda 100
sardine, Pacific 235
Sardinops sagax 235
saury, Atlantic 230–231
    eastern Pacific 231
    northern Pacific 230, 231
    southern 230, 231
scad, mackerel **142**
scalloped hammerhead shark 190–194, 191
scaly mackerel 108, **121**
school mackerel 120
Scomber australasicus 106,

    colias 106
    japonicus 106
    scombrus 106
Scomberesocidae 234
Scomberesox saurus saurus 231
    saurus scombroides 231
Scomberoides commersonnianus 145
    lysan 145
    tala 145
    tol 145
Scomberomorus cavalla 116
    commerson 114
    maculates 117
    munroi 120
    queenslandicus 120
    regalis 118
Scombridae 82, 100, 104, 106, 108, 111, 114, 121, 142
Sectator ocyurus 251
Seriola dumerili 123, 126, 128, 166
    hippos 123, 128
    lalandi 123, 126, 128, 166
    rivoliana 127
shark, basking 5, 214, 216, **217**, 217–218
    blacktip 195
    blue 3, 178, 195, **208**, 208–210
    bull 5, **202**, 202–203
    cookiecutter **212**, 212–213, 241
    crocodile **211**
    dusky 5, **199**
    great white 173, 173–176, 204, 207, 217
    hammerhead 190–194
    great 190–192, **191**
    scalloped 190–194, **191**
    smooth 190–192, **191**
    mackerel 18, 21, 173, 174, 181, 182, 184, 187, 211
    mako, longfin 3, **177**, 177–178, 181
    shortfin 19, 174, **177**, 177–181
    oceanic whitetip 3, 4,

ACKNOWLEDGMENTS

This book has had a long gestation period and many people have helped in a host of small and not-so-small ways over a number of years. It is likely that I will forget to acknowledge some of them, in which case, I apologize in advance.

I owe a great debt to the worldwide community of pelagic fish biologists for not only their open exchange of information and ideas but also their companionship. Izumi Nakamura (Kyoto University) was the first to introduce me to the extraordinary billfishes and has been a valued mentor throughout my career. Others who have helped in many ways include Kim Holland and David Itano (University of Hawaii), John Graves and Rich Brill (Virginia Institute of Marine Science), Michael Domeier (Marine Science Conservation Institute), John Gunn, John Stevens, Jock Young, Barry Bruce and Peter Last (all of CSIRO), Eric Prince, Bruce Collette, Dave Holts and John Hoolihan (United States National Marine Fisheries Service), Barbara Block (Stanford University), Alain Fonteneau (Institut de Recherche pour le Développement), Kurt Schaeffer (Inter American Tropical Tuna Commission), Chi-lu Sun (National Taiwan University), Mark McGrouther (Australian Museum), John Holdsworth (Bluewater Research), Tony Lewis, John Hampton and Don Bromhead (Secretariat of the Pacific Community), Peter Ward (Bureau of Rural Sciences), John Diplock (Hamata Pty Ltd), Kerstin Fritsches (University of Queensland) and Foss Leach (formerly, Te Papa, New Zealand). I am also grateful for the helpful comments of a number of anonymous reviewers of various sections of the book during its development.

Special mention needs to be made of illustrator Guy Harvey. When I first mentioned the project to Guy, he immediately, and generously, offered to paint all of the species selected. I am not sure he realized the enormity of the task, but he delivered all of the artwork with style and good grace, as is his nature. With the keen eye of a scientist coupled with countless hours observing these fishes below and above the water, Guy has been able to produce illustrations which truly capture the essence of the live fishes complete with their transient colors only really seen at sea. I would also like to acknowledge the early assistance from marine artists Craig Smith and Bernard Yau. While their work does not grace these pages, it helped considerably in conceptualizing the book for potential publishers.

Many photographers provided images and they are acknowledged alongside each photo. However, I am especially indebted to Bill Boyce (boyceimages.com), not only for the use of his wonderful photos but also for making me laugh a lot.

Special thanks also to the late Bob Dunn for help in obtaining much of the historic information and material used, and to Andrew Isles (Andrew Isles Natural History Books) for his sage advice and encouragement.

I have been indeed fortunate to spend many productive days at sea with a host of expert captains and fishers. They are too numerous to list, but I would like to especially mention Peter Bristow, John Dunphy, Tim Choate, Laurie Wright, John Johnston, Mike Sinclair, Tim Simpson, Neil Patrick and the late Peter Goadby and John O'Brien for their particular help and insights.

Copious thanks go to my agent Sally Bird (Calidris Publishing Services) and to my first publisher at UNSW Press, John Elliot, both of whom exhibited persistence and patience above and beyond the call of duty. More recently, Elspeth Menzies (UNSW Press) took on the task of bringing the project to fruition, which she managed with aplomb. Thanks also to Stephen Roche for a great editing job, and thanks especially to Josephine Pajor-Markus and Di Quick, both of UNSW Press, who have designed the book I imagined a long time ago.

To my sister, Jennifer Isaacs, author extraordinaire, thanks for being a great supporter and adviser.

Finally, to my partner Loani Prior, thank you for believing in me and my dream and for helping to make it happen.

This book is dedicated to my father, Geoffrey Pepperell. He fostered my early interest in marine life and was there when I caught my first fish. Thanks, Dad.